Functional Skills Maths Level 2

By Vali Nasser

Copyright © 2015

E-book editions may also be available for this title. For more information email: valinasser@gmail.com

All rights reserved by the author. No part of this publication can be reproduced, stored in a retrieval system, or transmitted in any form or by any means, electronic, mechanical, photocopying, recording or otherwise, without the prior permission of the publisher and/or author.

ISBN-13: 978-1511778565

ISBN-10: 1511778563

Edited version: 28/6/2015

The author will also do his best to review, revise and update this material periodically as necessary. However, neither the author nor the publisher can accept responsibility for loss or damage resulting from the material in this book

Table of Contents

INTRODUCTION 5

CHAPTER 1   ARITHMETIC PART I 7

Addition, Subtraction, Multiplication and Division 7

Subtraction 8

Division 10

Division 13

POSITIVE AND NEGATIVE NUMBERS 15

Examples: 15

Examples: 15

PRACTICE QUESTIONS 1 18

ANSWERS TO PRACTICE QUESTIONS 1 20

CHAPTER 2   ARITHMETIC PART 2 21

Time Based Questions 21

For converting time from 12 hour clock to 24 hour clock - see examples below 21

Rounding numbers and estimating 23

PRACTICE QUESTIONS 2 25

ANSWERS TO PRACTICE QUESTIONS 2 26

CHAPTER 3   ARITHMETIC PART 3 27

Questions involving percentages and fractions 28

PRACTICE QUESTIONS 3: 31

ANSWERS TO PRACTICE QUESTIONS 3 33

CHAPTER 4   ARITHMETIC PART 4 MORE FRACTIONS 35

Simplifying fractions 35

PRACTICE QUESTIONS 4 38

ANSWERS TO PRACTICE QUESTIONS 4 39

CHAPTER 5   PROPORTIONS AND RATIOS 40

Scales and ratios 41

Conversions 42

PRACTICE QUESTIONS 5 45

ANSWERS TO PRACTICE QUESTIONS 5 46

CHAPTER 6   SIMPLE INTEREST AND COMPOUND INTEREST 47

PRACTICE QUESTIONS 6 49

ANSWERS TO PRACTICE QUESTIONS 6 50

CHAPTER 7   FORMULAS 51

Formula 51

PRACTICE QUESTIONS 7 57

ANSWERS TO PRACTICE QUESTIONS 7 58

CHAPTER 8   REPRESENTING DATA IN TABLES 59

Tables 59

PRACTICE QUESTIONS 8 61

ANSWERS TO PRACTICE QUESTIONS 8 62

CHAPTER 9   SHAPES AND SPACES 63

PERIMETERS AND AREAS OF COMMON SHAPES 64

Perimeters, Areas and Volumes of common shapes 64

PRACTICE QUESTIONS 9 70

ANSWERS TO PRACTICE QUESTIONS 9 72

2D AND 3D SHAPES AND NETS 73

PLANS: 79

TESSELLATIONS 81

Other Tessellations **82**

PRACTICE QUESTIONS 10 83

ANSWERS TO PRACTICE QUESTIONS 10 84

CHAPTER 10   MORE DATA INTERPRETATION 85

Mean, Median and Mode **85**

Pie Charts **87**

Line graph **93**

PRACTICE QUESTIONS 11 95

ANSWERS TO PRACTICE QUESTIONS 11 97

CHAPTER 11   PROBABILITY 98

PRACTICE QUESTIONS 12 102

ANSWERS TO PRACTICE QUESTIONS 12 103

EXAM TYPE QUESTIONS 104

ANSWERS TO EXAM TYPE QUESTIONS 109

SOME USEFUL DEFINITIONS AND REMINDERS 113

Introduction

Functional Skills Maths Level 2 is aimed at helping you pass this exam with ease.

In the actual test, although the use of calculators is allowed, it is useful to do simple sums with confidence without using calculators. In addition to basic addition, subtraction, multiplication and division you are expected to be familiar with fractions, decimals, percentages, ratios and proportions in everyday context.

Everyday problems often involve being able to estimate as well as being able to work with simple formulas like Speed, Distance and Time and of course conversions from one type of currency to another when you go on holiday! Also, basic Data Interpretation or Statistics is useful to make sense of data that is presented numerically or visually in a workplace or in newspaper articles. There are also chapters on basic shapes and spaces since you also need to know how to work out lengths around a shape (perimeters), areas and volumes of basic shapes as well as be familiar with plans and drawing nets!

For example when tiling a bathroom you need to know the area you want to tile as well as the area of each tile. This way you can find out how many tiles you need to purchase. In essence you are expected to have skills to represent, analyse and interpret practical problems that you are likely to meet in everyday life.

Whichever Exam Board you are working for you will find this book useful for Functional Mathematics up to level 2.

Although the examples in this book start simply they gradually build up to Level 2.

These tests may be taken online and in the actual tests you can either pass or fail.

Although you are allowed to use calculators it is very easy to press the wrong key so it is good to check your answers by doing it again or step by step. If possible at least estimate the value of the answer. Then you will know whether the answer is sensible or not. This will help you to approach basic arithmetical problems with more confidence.

One thing to remember is that there is often more than one way of working out a given problem. It does not matter which method you use, so long as you feel comfortable with it.

About the Author

The author of this book has experience in both consultancy work and teaching.

The author's initial book 'Speed Mathematics Using the Vedic System' has a significant following and has been translated into Japanese and Chinese as well as German. In addition, his book 'Pass the QTS Numeracy Test with ease' is very popular with teacher trainees. Besides being a specialist mathematics teacher the author also has a degree in psychology. This has enabled him to work as an organizational development consultant giving him exposure to psychometric testing particularly applicable to numerical reasoning. Besides working in consultancy he also managed the QTS numeracy tests for teacher trainees at OCR in conjunction with the teaching agency. Subsequently he has tutored and taught mathematics and statistics in schools as well as in adult education.

He hopes that this book 'Functional Maths Level 2' will help those aspiring to pass this test which ever exam board you are studying for.

Chapter 1   Arithmetic part I

Addition, Subtraction, Multiplication and Division

Remember you can always use a calculator in the exam.

Example 1:

Nadia and Jane go to see a film. They also buy a carton of popcorn each.

The prices are given below. How much do they spend in total together?

Price of cinema ticket per person is £4
Price of each carton of popcorn is: £2

Method: The amount they each spend is £4 + £2 = £6

So the total Nadia and Jane spend = £6 + £6 = £12

Example 2:

Fatima buys a handbag for £15. She also buys some perfume for £3.50. How much does she spend in total?

Method: Fatima spends £15 + £3.50 = £18.50 altogether

(If you are doing this by a calculator make sure to put the decimal point after '3' in 3.50)

Consider another example using a Speed Method by compensating or adjusting.

Example 2:

A customer buys two items from a shoe shop, items A and B

The selling prices are as follows:

Item A costs £20.90
Item B costs £30.80

Find the total amount the customer has to pay.

Method: When you use a calculator don't forget to put the decimal point. (However, you might want to look at a quick method of achieving the same result as shown below):

Total cost = £20.90 + £30.80 = £21 - 10p + £31 - 20p = £21 + £31 – 10p – 20p = £52 -30p = £51. 70

Note: This time we have used the compensating or adjusting method as shown below:

£ 20.90 = £21 – 10p

£30.80 = £ 31 – 20p

Subtraction

Example 1:

Work out: 241 – 28

Method: Using a calculator we find that 241 – 28 = 213

Example 2: Anne has £40.50 left to spend for the week. She spends £38.75. How much does she have left?

Method: £40.50 - £38.75 = £1.75

You can check this by adding £1.75 to £38.75 to get the original amount which is £40.50.

Example 3:

At a pharmaceutical company a scientist has 10000 Milliliters of a particular liquid which she uses for her experiments. She uses up 8743 Milliliters after several experimental tests. How much does she have left?

Method: Using a calculator we find that: 1 0 0 0 0 – 8743 = 1257.

This means she has 1257 Milliliters of liquid left. (Again always check your answers) Check: If you add 1257 to 8743 you should get 10000

Multiplication of whole numbers by 10, 100, 1000.

It is useful to know that when you multiply a **whole number** by **10** you just add 1 zero to the number at the end. When you multiply a **whole number** by a **100** you add two zeros at the end and so on. Some examples below will help you understand this more clearly. Multiplying by 10 means the answer is 10 times bigger, similarly multiplying by 100 means the answer is 100 times bigger.

Examples:

(1) 45 × 10 = 450 (**add 1 zero to 45**)

(2) 67 × 100= 6700 (**add 2 zeros to 67**)

(3) 65 × 1000= 65000 (**add 3 zeros to 65**)

Example: A container for eggs has 6 eggs in it. I buy 10 such containers. How many eggs do I have altogether?

Method: Add '0' to 6 to get 60 eggs altogether. You can of course use a calculator by multiplying 6 by 10!

Decimal points.

Look at the number line below.

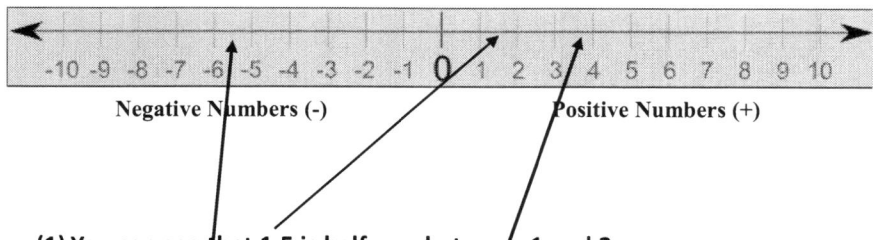

Negative Numbers (-)   Positive Numbers (+)

(1) You can see that 1.5 is half way between 1 and 2.
(2) Similarly 3.6 is more than three but less than 4.
(3) If you have a number like -5.34 it might be difficult to show but you know it's roughly here.

When adding, subtracting, dividing or multiplying don't forget to put the decimal point in the calculator.

Rules for numbers with decimals when multiplying by 10

When multiplying by 10, 100, 1000 move the decimal place the appropriate number of places to the right. **You can always use a calculator if you prefer!**

(1) 67.5 × 10 = 675 (the decimal point is moved 1 place to the right to give us 675.0 which is the same as 675)

(2) 67.5 × 100 = 6750 (this time move the decimal point two places to the right to give 6750.0 which is the same as 6750)

(3) 6.87 ×1000 = 6870 (in this case move the decimal point three places to the right to give the required answer.)

Division

Now consider examples involving division by 10, 100 and 1000 and other powers of ten.

(1) 450 ÷ 10 = 45 (You simply remove one zero from the number)

(2) 5600 ÷ 100 = 56 (This time you remove two zeros from the number)

(3) 45 ÷ 100 = 0.45 (No zeros to remove – so this time move the decimal point two places to the left to give us 0.45)

(4) 345.78 ÷ 100 = 3.4578 (Again simply move the decimal point 2 places to the left to give the answer)

(5) 456.78 ÷ 1000 = 0.45678 (Move the decimal point 3 places to the left as shown)

Questions involving powers of 10 (Use a calculator or use the 'speed method' whichever you prefer)

(1) Divide 27000 Milliliters by 10

(2) What is 7887 multiplied by 100?

(3) What is 67 divided by 100?

The answers are:

(1) 2700 ml (2) 788700 (3) 0.67

Also note, there is a short hand way of writing 100, 1000, 10,000 and larger powers of 10.

100 = 10^2 (10 squared, which is 10 × 10) = one hundred

1000 = 10^3 (10 cubed which is 10 × 10 × 10) = one thousand

10,000 = 10^4 ((10 to the power 4, which is 10 × 10 × 10 × 10) = ten thousand

1000,000 = 10^6 (10 to the power 6 which is 10 × 10 × 10 × 10 × 10 × 10) = one million

Higher powers can be written similarly.

Some other Multiplication Examples

Example 1: work out 15×13

Using a calculator we find that 15×13 = 195

Example 2: In a certain company 54 insurance agents manage to sell 14 insurance policies each in a particular month. How many insurance policies did these agents sell altogether in that month?

Method: Using a calculator we find the total insurance policies sold were 54 × 14 = 756

Multiplying by 5 quickly (Just in case you are interested to know)

Multiply the number by 10 and halve the answer.

Example 1: 5 × 4 = half of 10 × 4 = half of 40 = 20

Example 2: 5 × 16 = half of 10 × 16 = half of 160 = 80

Example 3: 5 × 23 = half of 10 × 23 = half of 230 = 115

Square numbers

We have already seen that 10^2 (10 squared) means 10×10 = 100

Similarly 5^2 = 5×5 = 25, 7^2 = 7×7 = 49, 15×15 = 225 and so on.

Example 1: Square the number 25

Method: We simply multiply 25×25 which equals 625

Example 2: Find the value of 18^2

Method: 18^2 means 18×18 = 324

Summary: To square a number you simply multiply the number by itself.

The Order of Arithmetical Operations

Remembering the order in which you do arithmetical operations is very important.

The rule taught traditionally is that of **BIDMAS.**

The **BIDMAS** rule is as follows:

(1) Always work out the **Bracket(s)** first
(2) Then work out the **Indices** of a number (squares, cubes, square roots and so on)
(3) Now **Multiply** and **Divide**
(4) Finally do the **Addition** and **Subtraction**.

Example 1: 4 + 13(7 – 2) this means add 4 to 13× (7 – 2)

Do the brackets first so (7 – 2) = 5, then multiply 5 by 13 to get 65 and finally add 4 to get 69

Example 2: Work out 2 + 8×3

Do the multiplication before the addition

So 8×3 =24 and 2 + 24 = 26

Example 3: Work out (2×15) + (4×8)

Do the bits in the brackets first: So we have (2×15) = 30 and (4×8) = 32

Now add 30 to 32 to get 62.

Example 4: work out $3^2 \times 5 - 9$

(3^2 means 3×3 or 3 squared)

Work out the **square of 3 first**, then **multiply by 5** and finally **subtract 9** from the result. So we have 3×3 = 9,   9×5 = 45 and finally 45 - 9 = 36

Summary: When working out sums involving mixed operations (e.g. +, - , x and ÷) you need to work out the steps in stages using the BIDMAS rule: So to work out 8 +25 ×12. Do the multiplication first, 25×12 =300, write down 300 then add 8 to get the answer 308. (When using a calculator be careful that you follow the above rules otherwise you could get the wrong answer!)

Division

Dividing a number by 2 is a very useful skill, since if you can divide by 2, you can by halving it again divide by 4.

Dividing by 2 and 4

Simply halve the number to divide by 2

Halving the number again is the same as dividing by 4

Example 1: 28 ÷ 2 =14 (half of 28)

Example 2: 48 ÷ 4 =24 ÷ 2 = 12 (check using a calculator)

Example 3: 64 ÷ 4 =32 ÷ 2=16 (check using a calculator)

Dividing by 5

An easy way to do this is to multiply the number by 2 and divide by 10.

Example 1:   120 ÷ 5 = (120 × 2) ÷ 10 =240 ÷ 10 =24

Example 2:   127 ÷ 5= (127 × 2) ÷ 10=254 ÷ 10=25.4

Dividing by other numbers: It is best to use a calculator.

Question involving division

Example1: In one particular week in a restaurant a bonus of £67.50 is divided amongst three waiters. How much does each one get in that week?

Method: £67.5 ÷ 3 = £22.50 per waiter

(In the exam for questions like the one above it is best to use a calculator)

Positive and Negative Numbers

Negative and Positive Numbers using the number line

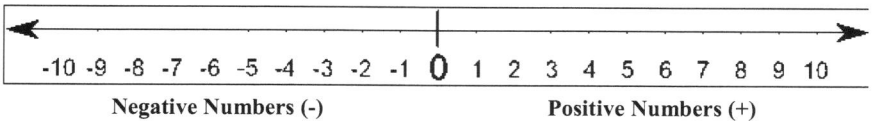
Negative Numbers (-)   **Positive Numbers (+)**

Numbers on the left decrease in value and numbers on the right increase in value.

Examples:

    (1) **3** is smaller than **5**
    (2) **-5** is smaller than **0**
    (3) **-9** is smaller than **–4**

<u>**Numbers on the right are bigger than numbers on the left**</u>.

Examples:

    (4) **9** is larger than **4**
    (5) **1** is larger than **–1**
    (6) **5** is larger than **–8**

You don't need to always use the number line. You can also use the method below:

When adding and subtracting positive and negative numbers it is worth knowing the following:

When you add two minus numbers you get a bigger minus number.

Example 1: $-4 - 6 = -10$

When you add a plus number and a minus number you get the sign corresponding to the bigger number as shown below:

Example 2: $+6 - 9 = -3$, whereas, $-6 + 9 = 3$

When you subtract a minus number from a number you need to note the results as shown below:

Example 3: $6 - (-3)$ we get $6+3 = 9$ (since $-(-3) = +3$)

Example 4: $7 - (+3)$ we get $7 - 3 = 4$ (since $-(+3) = -3$)

In this case note that $-(-) = +$. Also, $+(-) = -$ and $-(+) = -$.

Multiplying positive and negative numbers.

$(+) \times (+) = +$ (a plus number times a plus number gives us a plus number)

$(+) \times (-) = -$ (a plus number times a minus number gives us a minus number)

$(-) \times (+) = -$ (a minus number times a plus number gives us a minus number)

$(-) \times (-) = +$ (a minus number times a minus number gives us a plus number)

Dividing positive and negative numbers.

$(+) \div (+) = +$ (a plus number divided by a plus number gives us a plus number)

$(+) \div (-) = -$ (a plus number divided by a minus number gives us a minus number)

$(-) \div (+) = -$ (a minus number divided by a plus number gives us a minus number)

$(-) \div (-) = +$ (a minus number times a minus number gives us a plus number)

Summary: For both multiplication and division, like signs gives us a plus sign and unlike signs gives a minus sign

Example 1: $-£5 - £7 = -£12$ (Reason: If you give away £5 then give away £7, altogether you have given away £12. So you are £12 short or -£12

Example 2: Work out $2(4+6) - 21$

Work out the bracket first then times by 2 to get $2 \times 10 = 20$. Finally take away 21 from 20 get -1

Example 3: $3 + 13(7 - 2)$

The first part is 13 × 5 (we do the brackets and then multiply by 13)

13×5 = 65

Now we simply add 3 to 65 to get 68 as the final answer.

Example 4: Work out -4×-3

Method: minus times minus = plus. So -4×-3 = +12 or 12.

Example 5: Work out -6×2

Method: Minus times plus = minus. So -6×2 = -12

<u>**Summary**</u>

When using a calculator be careful in putting the right sign before the number.

(So when working out -5 + 3 put in -5 first then + 3, to get the answer)

Practice Questions 1

(1) 380 people work in a certain office in Bond Street. 290 people work in the office next door there. How many people in total work in both offices?

(2) There are 960 pupils in a certain comprehensive school. 470 of the pupils there are girls. How many boys are there?

(3) If I buy one ink cartridge for my printer it costs me £18.50. If I buy two together I get a discount of £7 in total. I decide to buy two ink cartridges. How much do I pay in total?

(4) John earns £11 per hour doing general maintenance work. The amount of work fluctuates. One good week he works for 43 hours. How much does he earn in this particular week?

(5) Fiona is having a small party. She bakes a cake which contains 8 ounces of sugar. She cuts the cake into 16 parts. How many ounces of sugar does each part contain?

(6) What is 300 divided by 6?

(7) What is 4000 divided by 100?

(8) How many sevens are there in 147?

(9) John has 400 kgms of cement which he wants to divide amongst 10 workers. How much does each worker get?

(10) Elizabeth buys some flowers. Each bunch of mixed flowers costs £3.50. Elizabeth buys 6 bunches of mixed flowers. How much does she pay in total?

(11) 3 supply teachers work at a school in a given week. They get paid £130 per day. 2 of the supply teachers work a full week of 5 days and the third supply teacher works for 3 days. What is the total amount paid for the 3 supply teachers that week?

(12) Which is bigger -8 or 9?

(13) The maximum temperature in Birmingham one December was 11 degrees centigrade and the minimum temperature was -2 degrees centigrade. What was the difference between the highest and lowest temperature?

(14) Arrange the numbers -3, -5, 0, -8, 1, 8, -7 from smallest to biggest.

(15) Multiply (a) -6 ×-7 and (b) -13×12

Answers to Practice Questions 1

(1) 670 people in total

(2) 490 are boys

(3) £30

(4) £473

(5) 0.5 ounces

(6) 50

(7) 40

(8) 21

(9) 40 kg

(10) £21

(11) £1690

(12) 9

(13) 13°C

(14) -8, -7, -5. -3, 0, 1, 8

(15) (a) 42   (b) -156

Chapter 2   Arithmetic Part 2

Time Based Questions

For converting time from 12 hour clock to 24 hour clock - see examples below

12 –Hour Clock	24 –Hour Clock
8.45 am	08:45
11.30 am	11:30
12.20pm	12:20
2.35 pm	14: 35 (after 12pm add the appropriate minutes and hours to 12 hours, in this case 2hrs 35mins +12hrs = 14:35)
8.45 pm	20:45 (8hrs 45mins + 12hrs = 20:45)
11.47pm	23:47 (11hrs 47mins +12hrs = 23:47)

The Convention is that if the time is in 24-hr clock there is no need to put hours after the time.

Also remember: 2.5 hours = 2 hours 30minutes (0.5 hours = half of 60 minutes)

2.25 hours = Two and a quarter hours = 2hrs 15 minutes

2.4 hours = 2 hours 24 minutes (0.4 hours = 0.4X60 = 24 minutes)

2.1 hours = 2 hours 6 minutes (0.1hours = 0.1 X 60 = 6 minutes)

Remember there are 60 minutes in one hour so 0.4 hours = 0.4×60 =24 minutes.

For other time based questions e.g. years, months, days, hours, minutes or seconds remember the appropriate units.

Example 1: At a company new candidates are mentored once a week for 12 minutes each. There are 15 candidates who are being mentored. The session starts at 11.30am. When does it finish? Give your answer using the 24 hour clock

Method: Clearly we need to first work out the total time it takes for all the candidates. Total time for 15 candidates is 15 × 12 = 180 minutes = 3 hours. So the mentoring session ends 3hrs after 11.30am – this means it ends at 2.30pm. However using the 24 hour clock the times it ends is 14:30

Example 2: Peter completes a lap in 2.3 minutes. How many minutes and seconds is this?

Convert 0.3 minutes into seconds. Since one whole minute = 60 seconds, then 0.3 minutes = 0.3X60 = 18 seconds. Hence Peter completes the lap in 2 minutes and 18 seconds.

(Note that 0.3 X 60 is the same as 3 X 6, hence this is equivalent to 18)

General Multiplication questions

Example 1 There are 4 medium size boxes containing 18 black jumpers each and 3 bigger boxes containing 23 black jumpers each. How many black jumpers are there altogether?

Method: 4 boxes of 18 each means there are 4 X 18 = 72 black jumpers

Similarly, 3 boxes of 23 each means, 3 X 23 =69 black jumpers

Adding all the black jumpers we get 72 + 69 = 141

There are a total of 141 black jumpers altogether.

Example 2

I buy 5 books for £3.97 each. How much change do I get from a £20 note?

Method: 5×3.97 = £19.85. Now subtract £19.85 from £20 and you can see that I get 15p change from my £20 note.

(Since 20 – 19.85 = 15)

Rounding numbers and estimating

We will start simply with rounding numbers to the nearest 10 and 100

Consider the number 271

Rounded to the nearest 10 this number is 270

Rounded to the nearest 100 this number is 300

(The principle is that if the right hand digit is lower than 5 you drop this number and replace it by 0. Conversely if the number is 5 or more drop that digit and add 1 to the left)

Try a few more:

5382 to the nearest 10 is 5380

5382 to the nearest hundred is 5400

5382 to the nearest 1000 is 5000

This rule can also be applied to decimal numbers:

3.7653 rounded to the nearest thousandth is 3.765

3.7653 rounded to the nearest hundredth is 3.77

3.7653 rounded to the nearest tenth is 3.8

3.7653 rounded to the nearest unit is 4

Tip: Remember to use common sense when rounding in real life situations:

Example: A book store wants to keep 120 books in the same size boxes. They can fit 22 books in a box. How many boxes will they need?

Method: Number of boxes required will be 120÷22= 5.5 (to one decimal place). But clearly, they cannot have 5.5 boxes. **So they need to have 6 boxes**

Estimating calculations quickly

Example 1: Work out (2.2 × 7.12)/4.12

We can quickly estimate that this is roughly equal to (2 X 7)/4 =14/4 which is around 3.5 or 4 rounded to the nearest unit. The actual answer is: 3.8 (to 1 decimal place)

Example 2: Work out 38 × 2.9 × 0.53

We can approximate 38 to be 40 to the nearest ten

We can approximate 2.9 o 3 to the nearest unit

We can approximate 0.53 to 0.5 to the nearest tenth

So the magnitude of the answer is 40 × 3 × 0.5

This is 120 × 0.5 =60 (approximately)

Practice Questions 2

(1) Work out 123×24

(2) What is 159 ÷3?

(3) Round the number 3.167 to two decimal places

(4) There are 5 boxes containing 19 pairs of shoes each and 3 bigger boxes containing 27 pairs of shoes each. How many pairs of shoes are there altogether?

(5) I buy 5 plates for £1.98 each. How much change do I get from a £10 note?

(6) A teacher wants to keep 140 books in the same size boxes. She can fit 18 books in a box. How many boxes will she need?

(7) In a workplace new apprentices are trained once a week for 25 minutes each. There are 6 apprentices being trained. The session starts at 10.30am. When does it finish? Give your answer using the 24 hour clock.

(8) John completes his morning jog in 12.4 minutes. How many minutes and seconds is this?

(9) Round the number 238 to the nearest 10.

(10) Anne manages to save £14.50 per week. How much will she save in one year?

Answers to Practice Questions 2

(1) 2952

(2) 53

(3) 3.17

(4) 176

(5) 10p

(6) 8 boxes (Since you cannot have 7.78 boxes!)

(7) 13:00

(8) 12 minutes 24 seconds

(9) 240

(10) £754

Chapter 3 Arithmetic Part 3

Fractions, decimals and percentages

A fraction is a part of a whole. So if there are 5 parts altogether then 2 parts can be expressed as 2 out of 5 or $\frac{2}{5}$. Similarly 3 parts out of 10 can be written as $\frac{3}{10}$. The top number is the number of parts you are interested in and the bottom number is the total number of parts.

Example 1: I cut a cake into 9 slices and eat two of them. What fraction of the cake have I eaten?

Method: There are 9 parts in the whole cake. I eat 2 parts out of 9. This means the fraction is $\frac{2}{9}$

I am sure most of you are aware that 'half' can be written as $\frac{1}{2}$ = 0.5. **This in turn is equal to 50%.** To show this is true put '1' in a calculator and divide (÷) by '2'. **You will get 0.5**

To find the percentage multiply 0.5 by 100. Remember to put the % sign after 50. This should give you 50%. (**Remember percent (%) simply means out of 100. So 50% means 50 out of 100, 25% means 25 out of 100 and so on**)

<u>Try and remember the following equivalences if you have forgotten them.</u>

Fractions, decimals and percentage equivalents

Fractions	Decimal	Percentage
$\frac{1}{2}$	0.5	50%
$\frac{1}{4}$	0.25	25%
$\frac{3}{4}$	0.75	75%
$\frac{1}{10}$	0.1	10%

Summary:

It is useful to remember the following equivalences if you can. Otherwise you can always use a calculator!

$\frac{1}{2}$=0.5=50%, $\frac{1}{4}$=0.25 =25%, $\frac{3}{4}$ =0.75 =75%, $\frac{1}{10}$=0.1=10%. To convert a fraction into a percentage, simply multiply the fraction by 100. **So to work out $\frac{3}{8}$ as a percentage. You put '3' in a calculator, divide by '8' and finally multiply by 100 and add a % sign to give you 37.5%.**

Questions involving percentages and fractions

Example 1: Find 25% of £250

Method: Find 50% of £250 and halve it again.

Half of £250 = £125, Half of £125 =£62.50, so 25% of £250 = £62.50

Using a calculator to do this is shown below:

Remember % (percent) means out of 100. **So 25% means** $\frac{25}{100}$ **= 0.25.** Now simply **multiply 0.25 by £250 which gives you £62.50**

Note: In the sentence '25% of £250' the word **'of'** corresponds to **'×'** (times).

Example 2: In a marketing department of 25 people there are 12 women and the rest are men.

(1) What fraction of the marketing department consists of men?

(2) What percentage is this?

Method: (1) Since there are 12 women, there are 13 men out of 25. So the fraction of men is $\frac{13}{25}$

(2) The percentage of men is $\frac{13}{25}$ × 100 = 52%, (Using a calculator **divide 13 by 25 first then multiply by 100 to get 52%**)

Example 3: 30% of the applicants for a certain job are male. There are 30 candidates in total. How many of the applicants are female?

Method: If 30% of the applicants are male, this means 70% (100% - 30%) are female. So we need to find 70% of 30 candidates. To find 70% of 30 we need to work out $\frac{70}{100} \times 30 = 21$. So 21 of the applicants are female.

Working out increase or decrease in percentages from original value

Example1: In a certain corner shop 16 packs of cereal A were sold in week 1. In the same shop 20 packs of the same cereal were sold in week 2. What was the percentage increase in the cereal packs A sold from week 1 to week 2?

Method: Increase in number of cereal packs A = 20 – 16 = 4. Original number of cereal packs = 16. The increase of 4 was based on 16 cereal packs. To work out the percentage increase we simply divide the increase by the original number of cereal packs and multiply this by 100. That is $\frac{4}{16} \times 100 = 25\%$

To work out decrease in percentages (uses the same principle as above)

Example2: The original price of a projector was: £150, the new price is reduced to £135. What is the percentage decrease in price? The decrease in price is £150 - £135 = £15. The decrease over the original price is $\frac{15}{150}$. To turn this into a percentage we multiply $\frac{15}{150} \times 100 = 10\%$. So the decrease in percentage price is 10%.

The basic formula to work out increase or decrease percentage change is shown below:

$$\frac{difference\ between\ final\ and\ original\ value}{original\ value} \times 100$$

Miscellaneous questions involving fractions and percentages

Example 1: Finding fraction of an amount

(1) Find $\frac{3}{4}$ of £600.

To do this, we first write the above question as $\frac{3}{4}$ × £600. (**Put '3' in a calculator divide by 4 and multiply the answer by 600. The answer is 450. Since the units are £, the actual answer is £450**

Example 2: Finding a fraction and turning it into a percentage

There are 40 builders in a small town in Yorkshire. In a particular month 5 builders are without work.

What is the percentage of builders that do not have work in this month in this small town?

The fraction of builders without work = $\frac{5}{40}$. To convert this to a percentage simply multiply the result $\frac{5}{40}$ by 100. So $\frac{5}{40}$ ×100 = **12.5%**

Practice Questions 3:

(1) Find $\frac{3}{4}$ of £700

(2) Find $\frac{4}{5}$ of £650

(3) Convert $\frac{4}{5}$ to a percentage

(4) What is $\frac{3}{5}$ as a percentage?

(5) What is 16^2 ?

(6) Indira and Mukesh decide to get married. Mukesh buys a ring for Indira which is priced at £750. However, he manages to negotiate a 20% discount. How much does Mukesh pay for the ring?

(7) Elizabeth and Gail go for a holiday to Malta for 5 days. They decide to go on an all - inclusive package at a price of £300 each. They manage to book on a last minute deal online which gives them a 30% discount each. How much do they pay in total?

(8) Fiona and Jacob decide to get a small terrace property in Liverpool. Fiona earns £22,500 per annum and Jacob earns £19,500 per annum. They visit their local Building society and are told they can raise a mortgage of 3 times their combined salaries. How much mortgage can they raise between them?

(9) Andy is thinking of purchasing a flat for £95000. He requires a deposit of 15%. How much does he need to have as a deposit?

(10) In an engineering factory workforce of 375 people there are 12 women and the rest are men. (a) What fraction of the workforce consists of men? (b) What percentage is this?

(11) What is $\frac{1}{2}$ of £370?

(12) What is 25% of £600?

(13) Sarah buys a dress on ebay for £20. A week later she manages to sell it for £28. What percentage profit does she make?

(14) Macy sells her old car at a 30% loss from the original price. She paid £2300 originally. How much loss did she make?

(15) A sweater is priced at £36. But there is a clearance sale and I can buy it at a quarter of the price. How much do I pay for the sweater?

Answers to Practice Questions 3

(1) £525

(2) £520

(3) 80%

(4) 60%

(5) 256

(6) £600

(7) £420 in total

(8) £126,000 Mortgage available

(9) Deposit required is £14,250

(10) (a) $\frac{363}{375}$ (b) 96.8%

(11) £185

(12) £150

(13) 40%

(14) £690 loss

(15) £9

Chapter 4   Arithmetic Part 4   More Fractions

Simplifying fractions

Reducing a fraction to its simplest form.

You can do this two ways (1) **you can divide the top number by the bottom number and the answer will be expressed as a decimal.**

Example1: Reduce $\frac{200}{500}$ to its simplest decimal form. If you convert this fraction into a decimal using a calculator you get 0.4. (Simply put 200 into a calculator and divide by 500)

Example 2

Reduce $\frac{70}{120}$ to its simplest form. If you divide top and bottom by 10 you get $\frac{7}{12}$

Another example:

Example 3: Simplify $\frac{2}{4}$

Method: You can probably see that $\frac{2}{4}$ is the same as $\frac{1}{2}$. This can be worked out as follows: Divide top and bottom of the fraction $\frac{2}{4}$ by 2. This means the top number becomes '1' and the bottom number becomes '2'. So a simpler way to write $\frac{2}{4}$ is $\frac{1}{2}$

Example 1:

60 people apply for a certain job vacancy. 12 people are short-listed for an interview. What is the proportion of people that are not short listed for the interview? Give your answer as a decimal.

Total number of people applying for this job = 60. Since 12 people are shortlisted, this means 48 are not shortlisted. Hence the proportion that is not shortlisted = 48/60. If you divide the top number by the bottom number you get 0.8. The answer as a decimal is 0.8

Example 2: Simplify $\frac{6}{9}$ in its lowest form expressing the answer as a fraction.

Method: Take the fraction $\frac{6}{9}$ and divide top and bottom by 3 to get $\frac{2}{3}$

Question involving mixed fractions

Example 1: Find $2\frac{3}{4}$ of £64

A number like $2\frac{3}{4}$ is called a mixed number because it consists of a whole number and a fraction.

To find the value of $2\frac{3}{4}$ of £64. **First put in the fractional bit into the calculator and then add it to the whole number.** So $2\frac{3}{4}$ = 3÷4 + 2 = **0.75 + 2 = 2.75.** Now we have to find 2.75 of £64. Remember 'of' means '×'. So 2.75 of 64 = **2.75×64 = £176**

Example 2: Find $1\frac{1}{5}$ of 55

Method: $1\frac{1}{5}$ = 0.2 + 1 =1.2. We now find 1.2 of 55 = 1.2 × 55 = 66

Example 3: Find $42÷1\frac{1}{2}$. Notice this time we have to **divide** 42 by $1\frac{1}{2}$.

Method: 42÷1.5 = 28

Example 4: I go to London from Birmingham and stop on the way for half an hour at Milton Keynes. The time it takes to get to Milton Keynes from Birmingham is $1\frac{1}{4}$ hours and after my break the time it takes to go to London from Milton Keynes is 1 hour. How long is the total journey including my break?

Method: Birmingham to Milton Keynes = $1\frac{1}{4}$ hours = 1.25 hours

Break for half an hour ($\frac{1}{2}$ hour) = 0.5 hours

Milton Keynes to London = 1 hour.

So the total time taken is 1.25 + 0.5 + 1 = 2.75 hours

(As there are 60 minutes in an hour 0.75 ×60 = 45 minutes.)

So another way of writing this answer is 2 hours 45 minutes.

Practice Questions 4

(1) Work out half of £560

(2) Karen bakes small ginger cakes for a party. For each of these cakes she needs 2 ounces of sugar. If she bakes 30 of these cakes how many ounces of sugar does need?

(3) Richard gets 25 days of holidays per year. He has already used $\frac{3}{5}$ of his holidays. How many days has he got left?

(4) What is $2\frac{2}{5}$ of £350?

(5) Write $3\frac{3}{4}$ as a proper fraction

(6) Helen has used up 5 gigabytes of her 8 gigabytes memory on her smart phone. What fraction of the memory does she have left?

(7) Elizabeth manages to book a £210 holiday online with $\frac{1}{3}$ of the price off. What price does she pay for the holiday?

(8) Ahmed has worked overtime for 16 hours this month. His normal pay is £8 per hour and if he works overtime he gets $1\frac{1}{2}$ times as much. How much overtime pay does Ahmed get this month?

(9) Which is bigger $\frac{3}{8}$ or $\frac{2}{7}$?

(10) What is $60 \div 3\frac{1}{2}$?

Answers to Practice Questions 4

(1) £280

(2) 60

(3) 10 days left

(4) £840

(5) $\frac{15}{4}$

(6) $\frac{3}{8}$

(7) £140

(8) £192

(9) $\frac{3}{8}$ is bigger

(10) 17.14

Chapter 5 — Proportions and ratios

Although proportion and ratio are related they are not the same thing – see example below for clarification.

Example: In a class there are 13 girls and 10 boys. The **ratio of girls to boys is** 13:10, and the **proportion of girls in the class** is 13 out of 23 or $\frac{13}{23}$ (Since the total number of pupils is 23, the bottom number is 23)

Questions based on proportions and ratios

Example 1

In a class of 17 pupils, 9 go home for lunch. What is the proportion of pupils in this class that have lunch at school?

Since 9 out of 17 pupils go home, this means 8 pupils have lunch at school.

As a proportion this is 8 out of 17 or $\frac{8}{17}$

Example 2: In a certain work place the ratio of males to females is 2: 3 there are 250 workers altogether. How many of these are male?

Step 1: Find out the total number of parts. You can do this by adding up the ratio parts together. E.g. 2:3 **means there are (2+3) = 5 parts altogether.** This means 1 part = one fifth of 250 workers = 50 workers. ($\frac{1}{5}\times 250 = 50$)

Since the ratio of male to female is 2:3, there are **2X50 males** and **3X50 females**

The number of males in this workplace =2X50 = 100

Example 3:

£100 is divided in the ratio 1: 4 how much is the bigger part?

The total number of parts that £100 is divided into is 5 (to find the number of parts simply add the numbers in the ratio, which in this case is 1 and 4)

Clearly, 1 part equals £20 (100 divided by 5), so 4 parts is equal to £80 (Since 4×20 =80). So £80 is the bigger part.

Example 4:

£1500 is divided in the ratio of 3 : 5 : 7

Find out how much the smallest part is worth?

Clearly £1500 is divided into a total of 15 Parts (add up the ratio parts 3 : 5 : 7)

So each part is worth £100 (£1500 divided by 15)

So 3 parts (this is the smallest part) equals 3 ×100 = £300

Example 5:

As we have seen, sometimes ratios are expressed in ways, which may not be the simplest form. Consider 5:10

(a) You can re-write 5:10 as 1:2 (divide both sides of 5:10 by 5)

(b) 4: 10 can be re-written as 2: 5 (divide both sides of 4:10 by 2)

(c) 15 : 36 simplifies to 5 : 12 (divide both sides of 15:36 by 3)

Example 6: A team of 10 people can deliver 6000 leaflets in a residential estate in 3 hours. How long does it take 6 people to deliver these leaflets?

Method: 1 person will take 10 times as long or 3X10 = 30hours

This means 6 people will take 30÷6 =5 hours

Scales and ratios

Consider that you are reading a map and the scale ratio is 1: 100000

This means for every one cm on the map the actual distance is 100000 cm or, put another way every one cm on the map, the distance = 1000 m (divide 100000 by 100. To get the result in metres) now, 1000 m = 1km (divide 1000 by 1000 to get 1 since 1km =1000m)

(Scales can also be used in other areas such as architectural drawings)

Question based on scales

I note that the map I am using has a scale of 1: 25000. The distance between the two places I am interested in is 12cm. What is the actual distance in km?

Method: 12 cm on the map corresponds to 12 x 25000 = 300x 1000 =300000cm

=3000m = 3km

(300000 /100 to convert to metres = 3000 m, now divide 3000 by 1000 to convert to km)

Hence the distance between the two places is 3km

Conversions

Conversions are often useful in changing currencies for example from pounds to dollars or euros and vice-versa. It is also useful to convert distances from miles to kilometres or weights from kilograms to pounds and so on.

Basically a conversion involves changing information from one unit of measurement to another. Consider some examples below:

Question based on conversions

Example 1:

I go to France with £150 and convert this into Euros at 1.2 Euros to a pound.

(1) How many Euros do I get? **(2)** I am left with 39 Euros when I get back home. The exchange rate remains the same. How many pounds do I get back?

Method: (1) Since 1 pound = 1.2 Euros, I get 150 X 1.2 =180 Euros in total.

(2) When I get back I change 39 Euros back into pounds. This time I need to divide 39 by 1.2

So 39÷1.2 =32.5. This means I get back £32.50

Example 2

The formula for changing kilometres to miles is given by:

$M = \frac{5}{8} \times K$. Use this formula to convert 68 kilometres to miles

Method: substitute **K** with 68 and multiply by $\frac{5}{8}$

This means $M = \frac{5}{8} \times 68$. Using a calculator this comes to 42.5 miles

It is worth reviewing some common Metric and Imperial Measures as shown below

Metric Measures

1000 Millilitres (ml) =1 Litre(l)

100 Centilitres (cl) =1 Litre (l)

10ml =1 cl

1 Centimetre (cm) =10 Millimetres (mm)

1 Metre (m) = 100 cm

1 Kilometre (km) =1000 m

1 Kilogram (kg) =1000 grams (g)

Imperial Measurements

1 foot =12 inches

1 yard =3 feet

1 pound = 16 ounces

1 stone =14 pounds (lb)

1 gallon = 8 pints

1 inch = 2.54 cm (approximately)

Question on conversions

Example 1: How many grams are there in 2.5kg?

Method: Each Kg = 1000g. So 2.5kg = 2.5×1000 g = 2500g

Example 2: How many cm are there in 0.5km?

Method: Since, 1metre = 100 cm and 1km = 1000m, this means 0.5km = 500m = 500×100 cm = 50000 cm

Example 3: Robert is going to the USA. He changes £120 into US dollars. The exchange rate is $1.65 to one pound sterling. How many dollars does Robert get?

Method: Since we know that one £ sterling = $1.65 this means £120 = $120×1.65 = $198

Example 4: A ramblers' group, go on a walking tour whilst in the South of France. They walk from Perpignan to Canet Plage which is approximately 11 km away. After a lunch break and some time on the beach, they walk back to Perpignan. How many miles in total do they walk on that day? (You are given that 8 km is approximately equal to 5 miles.) Give your answer as a decimal.

Method: Total distance walked = 22Km (11 + 11). To convert this into miles we have to multiply 22 by 5 and then divide by 8 (Since 8 km = 5 miles)

That is $22 \times \frac{5}{8}$ =13.75 miles (divide 5 by 8 and multiply the answer by 22)

Practice Questions 5

(1) Mary invites 28 people to her birthday party. 23 of them are girls. What is (a) the proportion of girls in the party? (b) What is the proportion of boys?

(2) Helen cycles to work 3 days out of 5 during a working week. What is the proportion of days she does not cycle to work?

(3) Ahmed has sandwiches 5 times a week for his lunch. For 2 days he has curry. What is the proportion of times he has sandwiches for lunch?

(4) The ratio of sugar to fibre in an orange is 1:7. Assuming an orange weighs 80 grams how many grams of sugar does it contain?

(5) £500 is split between John and James in the ratio of 2:3. How much does James get?

(6) £1200 is divided between 3 people in the ratio of 1:2:3. How much is the largest amount that a person gets?

(7) In 2014 in a particular school 79 people take French out of a total of 690 pupils. What is the proportion of pupils who take French in this school?

(8) I am travelling from Waterloo to Surbiton. The scale on the map is 1: 200000. I measure that on the map this distance is 12cm. How many kilometres is this?

(9) This time I use a map where the scale shown is 1cm = 1 mile. I measure the distance between two places and the distance on the map is 10 cm. How many kilometres is this if you are given that 8 km is equal to 5 miles.

Answers to Practice Questions 5

(1) (a) The proportion of girls are $\frac{23}{28}$
(b) The proportion of boys are $\frac{5}{28}$

(2) $\frac{2}{5}$

(3) $\frac{5}{7}$

(4) 10gms

(5) £300

(6) £600

(7) $\frac{79}{690}$

(8) 24km

(9) 16 km

Chapter 6   Simple Interest and Compound Interest

Simple Interest

Example 1: I have £5000 in a building society account which pays me simple interest of 3% per annum. I keep my money for 3 years. How much in total will I have at the end of the 3 year period?

Method: At the end of the first year the total interest I will receive is 3% of £5000. This is $\frac{3}{100} \times 5000 = \frac{15000}{100}$ = £150 per annum.

At the end of 3 years I will receive 3X £150 = £450 total interest. This means the total amount I will have is £5450 (original £5000 plus 3 years of simple interest)

You can if you like use the formula below to work out the total simple interest over a given period of time.

I =PRT where I = Total interest, P = Principal amount (original amount), R is the annual interest rate and T = the time in years.

So in the above case I =5000X $\frac{3}{100}$ X3 = £5000 × $\frac{9}{100}$ = £450

Finally, to find the total amount we have at the end of the 3 year period, we simply add £450 to the original £5000 to get £5450

Example2: Work out the final amount at the end of one year if there is a 10% increase per annum and I have £3000 to start with.

The traditional method is to work out 10% of £3000 first. Then add this answer to £3000 to get the final answer.

So 10% of £3000 = £300. So the final price after a 10% increase is £3000 + £300 =£3300

Here is fast and efficient method to work out the final price:

Simply work out 1.1× 3000

Since 1.1 denotes a 10% increase.

Why 1.1? Since 100% plus 10% =1 + 0.1 = 1.1

Now 1.1× 3000 = £3300 which is the final answer

Compound Interest (you can use a calculator to work out the examples below!)

Now consider a problem involving recurring percentage changes

Example 1

Find the value of £5000 if I gain a profit of 10% the first year followed by (10% of the new amount) in the second year. (This is called compound interest)

THIS MEANS THE INCREASE IS 1.1× FOLLOWED BY 1.1× AGAIN

Or $(1.1)^2$ × 5000

$(1.1)^2$ × 5000 = 1.21× 5000 = £6050

So the final value is £6050

Example 2

I buy a one bedroom apartment for £200,000. It increases in value by 5% per annum

How much will it be worth in 15 years?

Method: Increase after 1 year will be 1.05X£200,000, after two years it will be: $(1.05)^2$ ×£200,000, after three years it will be $(1.05)^3$ × $200,000. So, after 15 years it will be worth $(1.05)^{15}$ × £200,000 = £415786

Example 4:

A car depreciates by 30% per annum. I buy it at £18000. What is its value in 5 years' time? Give your answer to the nearest pound

Method:

After one year its value will decrease by 30%, so its new value will be 70% of original as shown below:

£18000×0.7, hence, after five years its value will be £18000× $(0.7)^5$

Value after five years is £3025

Practice Questions 6

(1) Anne has £2000 in an ISA savings account. The interest rate is 3% compound. (a) How much will Anne have in total at the end of one year? (b) How much will she have in total after 2 years? Show as much of your working out as possible.

(2) George and Wendy have a property which they bought 3 years ago for £85000. Property prices have risen by 6% per year. How much is their property worth now?

(3) Asif buys a second hand Ford Fiesta for £2500. After 3 years it has gone down in value by 55%. What price can he sell it for now?

(4) Peter has £6000 in a bank account which pays him simple interest at 2.5% per year. How much interest will he earn after 4 years?

(5) Mary has a buy to let property which gives her an income before expenses of 7% on the value of the property. The property is worth £90,000. (a) How much income before expenses does she earn in a year? (b) Her mortgage expenses are £3,600 and her management expenses for letting are £630. How much does she get after all expenses? Show your working.

Answers to Practice Questions 6

(1) (a) After one year she will have £2060 in total (b) After two years she will have £2121.80 in total

(2) £101,236.36

(3) £1225

(4) £600

(5) (a) Income she gets is £6300

(b) After expenses she gets £6300 - £3600 - £630 = £2070

Chapter 7   Formulas

Formula

A formula describes the relationship between two or more variables. Consider a simple case first.

Example 1: A company pays 40p per mile and certain meal expenses when their sales employees visit clients. The cost of claiming mileage is calculated using the formula given, payable at 40p per mile and a fixed cost of £25 for general expenses. The formula is given by **C = M× 0.4 + 25**, where **M** represents the number of miles travelled and **C** represents the total cost in pounds payable to the employee by the company.

So for example if an employee has to travel 40 miles from her home to the client, the employee can claim 80 miles altogether for the journey to the client and back + £25 as shown below by the formula.

Using the formula we have C = 0.4× 80 + 25 = 32 + 25 = £57

(Explanation of working out: Using BIDMAS we multiply before adding. So 0.4×80 =32, finally add 32 and 25 together to get 57)

Example 2:

(1) The formula for working out the distance depends on the speed and time taken in the appropriate units.

D = S×T where D is the distance, S the speed and T is the time.

What is the distance travelled if my speed is 60kmh and I travel for 1hour and 30 minutes. 1 hour 30 minutes corresponds to 1.5 hours so, using the formula, D = 60×1.5 = 90 km.

That is, the distance equals **90km**

(2) The formula for working out the speed is given as Speed= Distance/Time

That is **S = D÷T**

Work out the average speed with which I travel, if I cover 100 miles in 2.5 hours.

Since **S = D÷T**, this means S = 100÷2.5 =40 mph (Notice the units for the first example were in kilometers and units for the second example were in miles

(3) The formula for working out time taken is given by T = D÷S

Calculate the time taken to cover 90 miles if I travel at 60mph?

Time taken, T= D÷S, so T = 90÷60 = 9÷6 =3÷ 2 = 1.5 hours or 1 hour and 30 minutes.

Example 3

The formula for converting the temperature from Celsius to Fahrenheit is given by the formula: F= $\frac{9}{5}$C +32 (where C is the temperature in degrees Centigrade)

If the temperature is 10 degrees Celsius then what is the equivalent in temperature in Fahrenheit?

Using the formula F= $\frac{9}{5}$C +32, and substituting 10 in place of C, we have F= $\frac{9}{5}$ × 10 +32 = $\frac{90}{5}$ + 32 =18+32 =50. Hence, 10 degrees centigrade = 50 degrees Fahrenheit

(**Explanation of working out above**: There are no brackets to worry about. When working out $\frac{9}{5}$ X 10 +32, divide 9 by 5 then multiply by 10 to get 18, finally add 18 and 32 together to get 50.)

Example 4: Convert 68 degrees Fahrenheit to degrees Celsius. The formula for converting the temperature from Fahrenheit to Celsius is given by:

C= $\frac{5}{9}$(F-32). To change 68 degrees Fahrenheit to degrees Celsius we can substitute for F in the formula C= $\frac{5}{9}$(F-32)., C = C= $\frac{5}{9}$(68-32). =5 X 36/9 = 5 X 4 =20

Hence, 68 degrees Fahrenheit =20 degrees Celsius

(**Explanation of the working out above**: Using BIDMAS we work out the bracket first. This gives us 68-32 =36. We now divide this by 9 and multiply by 5. Clearly 36÷9 =4 and finally 5X4 =20)

We have seen that formulas can be important in conversion problems

Earlier we saw the formula: S = D ÷T, that is, Speed = $\frac{Distance}{Time}$. Sometimes in the On screen questions you may be shown a distance time graph for a school coach trip and asked to work out average speed for a particular part of the journey and the time the coach was stationary. See example below

Example: A school trip by coach to a heritage site leaves at 1200 hrs from the school. The coach arrives at the destination at 1300hrs. It then stops so the pupils can look around the site. Finally after looking around the site it leaves and arrives back at school at 15:30hrs. (1) How long did the coach stop for? (2) What was the average speed on the return journey?

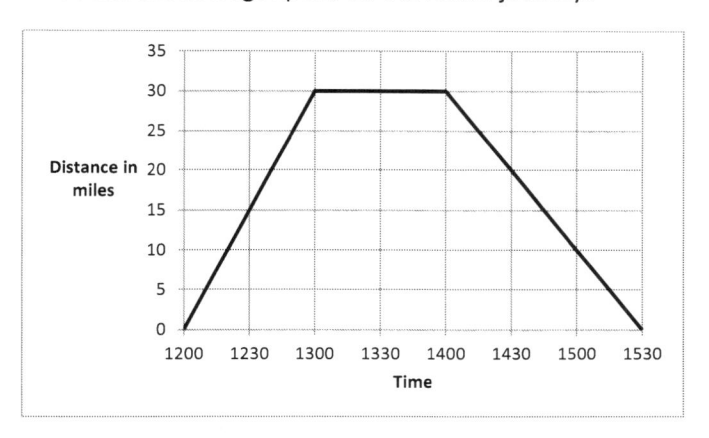

(1) From the distance-time graph above you can see it was stationary from 1300 – 1400hrs, which is 1hr

(Between these time intervals no further distance is covered, so it is stationary – see the vertical axis at 30 miles)

(2) The return journey starts at 1400hrs and ends at school at 1530hrs = 1.5 hrs.

Since **Speed** = $\frac{Distance}{Time}$, this means speed = 30÷1.5 = 20 mph

Formula for working out your body mass index.

Bodyweight in kilograms divided by height in metres squared

Example of another formula

BMI (body mass index) = weight in kgms÷$(height)^2$ or $\frac{W}{H^2}$

Example: work out the body mass index of someone who is 65kg in weight and is 1.5m tall.

Method: BMI = weight in kgms ÷ $(height)^2$

(You are given that BMI = $\frac{W}{H^2}$. Where W is the weight is in Kgms and H is the height in meters. Work out the BMI for a Hussein whose height is 1.5m and weight is 65 kg)

Method: Using the formula to work out the body mass index given: BMI = $\frac{W}{H^2}$ we get **BMI = 65÷ (1.5 × 1.5) = 22.86** (Remember to **square the height** and then **divide this result into 65**)

Changing the formula around – (change of subject of formula)

Example 1: You are given that S = A + B, Change the formula so A is on one side and everything else on the other side. (We can also say make A the subject)

Method: You are given that S= A + B. If you subtract B from both sides you get

S – B = A + B – B. This simplifies to S – B = A. This means **A = S – B**

(The basic rule is: Whatever you do to one side of the formula you do the same to the other side- in the example above we subtracted B from both sides)

Example 2: Given that **Speed** = $\frac{Distance}{Time}$, Make distance the subject of the formula.

Method: Multiply **both sides by** Time. So we get **Speed×Time** = $\frac{Distance \times Time}{Time}$

This means **Speed×Time = Distance** or re-writing this we have:

Distance = Speed×Time

Example 3: The velocity of a particle is given by the formula v = u + at, where 'v' is the final velocity, 'u' the initial velocity and 'a' the acceleration. Make t the subject of the formula.

Method: We are given that v = u + at.

Step1: Subtract u from both sides to get rid of 'u' from the right hand side. This leaves us with v – u = at.

Step 2: Now divide both sides by t to get rid of 't' from the right hand side so we get $\frac{v-u}{a}$ = a. Re-writing this we get **a** = $\frac{v-u}{a}$

Summary

(1) To get rid of +a you –a from both sides

(2) To get rid of – a you + a to both sides

(3) To get rid of '×' you divide (÷) both sides by the same letter

(4) To get rid of '÷' you multiply (×) both sides by the same letter

Remember the basic rule 'whatever you do to one side you do the same to the other' (So if you want to get rid of ×a you ÷ **both sides** by a)

So for example in the formula x = y + m. To make y the subject, we need to get rid of m, so you −m from both sides giving you x − m = y. That is y = x − m

You might find the following conversions useful to go through

1 km = 5/8 mile

1 mile = 8/5 km

Practice Questions 7

(1) You are given that Speed = $\frac{\text{Distance}}{\text{Time}}$. Calculate my speed in m.p.h if I travel 25 miles in half an hour.

(2) John is a travelling salesman. He gets expenses at 25p for every mile he covers and £25 to cover his lunch and dinner every day. The formula the company uses to work out his expenses for a given number of days is given by E = 0.25m + 25d (where m is the number of miles covered and d is the number of days he has been travelling). One week he travels 360 miles in 4 days. How much will he receive in expenses altogether for this week?

(3) Whilst she is in France Miranda wants to convert the number of kilometres she has covered into miles. She notes the formula for converting kilometres into miles is given by M = $\frac{5}{8}$ X K. Assuming she covers 480 km how many miles is this?

(4) The formula for converting degrees centigrade into Fahrenheit is given by the formula F= $\frac{9}{5}$C +32. What is the temperature in Fahrenheit if it is 25 degrees centigrade?

(5) Joe has been told that he needs to reduce his body mass index to below 25 as he is presently very overweight. He presently weighs 90kg and is 1.8m tall. He decides to lose 10kg in the next 3 months. The formula for working out the body mass index is given by:

BMI = Weight ÷ $height^2$

- (a) What is his present BMI?
- (b) What will be his BMI after losing 10Kg, give the answer to one decimal place?
- (c) Does he achieve his goals?

(6) Make m the subject of the formula from s = m + 2t

Answers to Practice Questions 7

(1) 50 mph

(2) E = 0.25m + 25d, so E = 0.25×360+ 25×4 = £190

(3) 300 miles

(4) 77°F

(5) (a) BMI = 27.8 (to one decimal place)

(b) After losing 10Kg Joes's BMI = 24.7 (to one decimal place)

(c) Yes

(6) m = s – 2t

Chapter 8 Representing data in tables

Tables

Tables such as the ones shown below can be used to represent different types of data. This in turn can help us to interpret and analyze the information given. The examples shown below demonstrate this.

Example 1:

This table shows different languages being studied in a certain college by boys and girls

	German	French	Polish	English	Total
Boys	10	10	5	20	45
Girls	5	15	5	25	50
Total	15	25	10	45	95

Typical questions

(1) How many girls are in the French class?

Method: Look **down the French column** and **across the Girls row** as shown by the small arrows. You can see that **15 girls study French** in this college.

(2) What is the total number of boys who study languages?

Method: Going across the boys row and then down the total column you can see that the total number of boys who study languages is 45.

Example 2:

From the train time table shown below when would I reach Lichfield if I took the 13.25 train from Birmingham?

Birmingham	13.05	13.15	13.25	13.35
Erdington	13.15	13.25	13.35	13.45
Sutton Coldfield	13.25	13.35	13.45	13.55
Lichfield	13.45	13.55	14.05	14.15

Method: Starting from Birmingham at 13.25 I would reach Lichfield at 14.05 as shown.

Example 3: (Slightly harder example)

		Method of Transport			
		Car	Bus	Walking	Other
Type of Schools	Inner City	28%	32%	24%	16%
	Suburban	62%	18%	12%	8%

From the data above you can see that 32% of children take the bus in Inner City schools compared to 18% who take the bus in suburban schools. Similarly, 62% of pupils in suburban schools arrive by car as compared to 28% in inner city schools. You can also compare other modes of transport between the two schools.

Example 4:

The distances between various towns on a journey from Birmingham to London are given below in miles. What is the distance between Coventry and Milton Keynes?

	Birmingham	Coventry	Milton Keynes	London
Birmingham	XXXXXXXXX	25	70	120
Coventry	25	XXXXXXXX	45	95
Milton Keynes	70	45	XXXXXXXX	50
London	120	95	50	XXXXXXXXX

Method: Look at the column under Coventry and row across Milton Keynes. They meet at the bit shown. The distance is between Coventry and Milton Keynes as per the table above is 45 miles.

Practice Questions 8

(1) Michelle is going to see a friend in Leeds. Michelle lives in London near Kings Cross. She looks at the train time table shown below and decides to take the 14.10 train. How long is her train journey?

Depart	Arrive
London Kings Cross	Leeds
13.10	16.49
13.42	17.01
14.10	17.49
14.42	18.02

(2) Five classes raise money for a charity. The amount raised per class as well as the number of pupils in each class is shown in the table below:

Class	Number of Pupils	Amount of money raised in £
A	27	37.50
B	22	32.75
C	28	40.20
D	21	22.50
E	24	24.80

(a) How much is raised by the pupils in class D?
(b) Which class raises the most money?

(3) A Deputy Head created the following table showing the number of pupils in each year group who had music lessons. How many pupils did not have music lessons in year 9?

Year Group	Number of pupils	Number of pupils who have music lessons
7	92	10
8	101	18
9	105	14
10	96	13
11	102	11

Answers to Practice Questions 8

(1) 3h 39m

(2) (a) £22.50

    (b) Class C

(3) 91 pupils do not have music lessons

Chapter 9   Shapes and Spaces

Some common shapes

Triangles

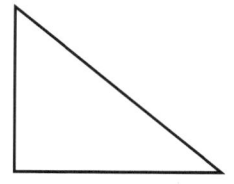

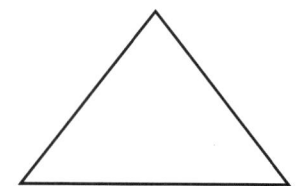

4 sided shapes

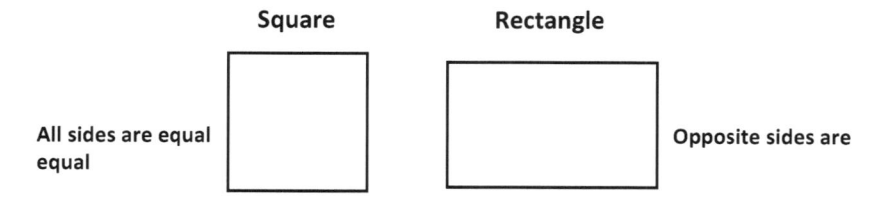

Square   **Rectangle**

All sides are equal equal   Opposite sides are

Perimeters and Areas of common shapes

Perimeters, Areas and Volumes of common shapes

Consider the shapes below:

(1) **Rectangle:** The shape of a rectangle is shown below:

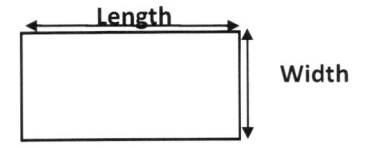

```
        Length
     ┌──────────┐
     │          │ Width
     └──────────┘
```

Perimeter of a rectangle = (distance around the outside of the rectangle) = 2×length + 2×width

Area of a rectangle = Length X Width

Note: Area is measured in units squared, e.g. cm^2 or m^2 and perimeter (distance all round a shape) is measured in the appropriate units e.g. cm or m

As explained above perimeter is simply the length around the outside of a shape measured in appropriate units. (Kilometres, metres, centimetres, millimetres, feet, inches, etc.)

Example 1: Find the perimeter of the rectangle whose length is 7 metres and width is 3 metres as shown below:

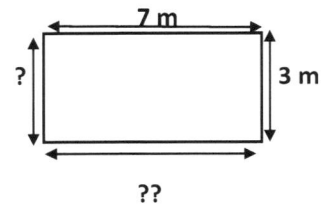

```
         7 m
     ┌──────────┐
   ? │          │ 3 m
     └──────────┘
         ??
```

Method: The perimeter all the way round the rectangle is 7m + 3m + 7m + 3m = 10m + 10m = 20m

(**Note:** Because the opposite sides of a rectangle are equal this means in the diagram above '?' = **3 meters** and '??' = **7 metres**)

Example 2: John wants to measure the perimeter of his garden. The shape and measures of the garden are shown below. What is the perimeter of the garden?

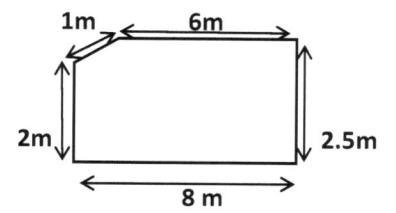

Method: To find the perimeter of the garden John simply has to add all the measures shown around the shape of the garden. So the perimeter is 1m + 6m + 2.5m + 8m + 2m = 19.5m

Areas

Area is simply the space inside a shape measured in square units.

Example 1: Find the area of the rectangle below:

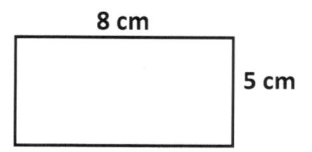

Method: Area of a rectangle is length × width. This means the area of the rectangle shown is 8×5 = 40 cms squared or 40 cm^2. (Remember area is measured in units squared.)

Triangle

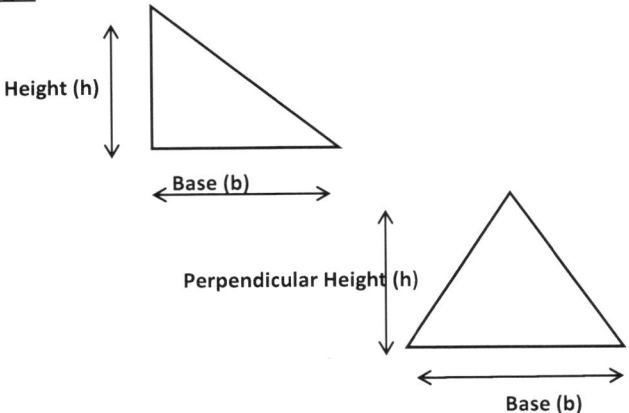

Area of a triangle = 1/2 × base × height or $\frac{b \times h}{2}$ (The height is the perpendicular height relative to the base. Perpendicular means it meets the base at right angles or 90°)

Example: Find the area of the triangle shown below:

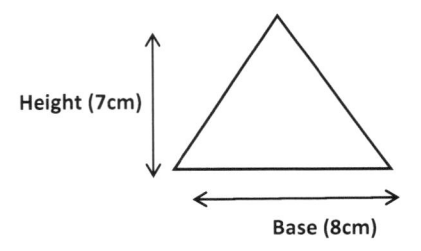

Method: Area of a triangle is given by: $\frac{1}{2}$× **base × height.**

Area = $\frac{1}{2}$ ×8×7 = **28** cm^2. So area of the triangle above is **28** cm^2

Circle

Diameter: A diameter is a line that goes **through the centre of a circle** and touches the **ends of a circle as shown**

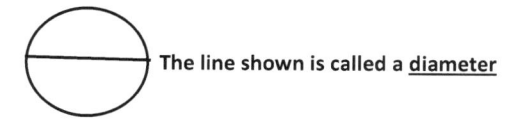

The line shown is called a <u>diameter</u>

Radius: This is simply half the length of the diameter

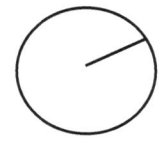

Circumference: This is another way of saying the perimeter of a circle. In other words it is the length all the way around the circle.

There is a special formula for working out the circumference.

Circumference = πd or $2\pi r$ (This means Circumference = π×diameter or 2×π×radius)

The value of π (**pronounced pi**) can be found by pressing the π key on the calculator.

If you want to impress your friends and remember pi to 6 decimal places you can use a simple memory trick by remembering a sentence like '**How I wish I could calculate pi'** = 3.141592 (Notice **How** =3, **I** =1, **wish** = 4, etc)

<u>However, you can use π= 3.14 as an approximate value</u>

Example: Work out the circumference of a circle whose radius is 5cm.

Circumference = $2\pi r$ =2×π×radius = 2×3.14×5 = 31.4cm to 1 decimal place.

Area of a circle is πr^2 (this means the value of π(pi) multiplied by radius squared)

Example: Work out the area of the circle below whose diameter is 20 cm.

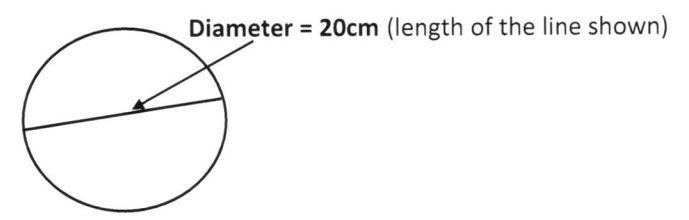

Diameter = 20cm (length of the line shown)

Method: To find the area we first we need to find the radius. The radius is half the diameter. This means the radius = half of 20 cm = 10 cm.

Area = πr^2 = 3.14×10×10 = 314 cm^2

So area of this circle is 314 cm^2

Volume of a cuboid (box shaped)

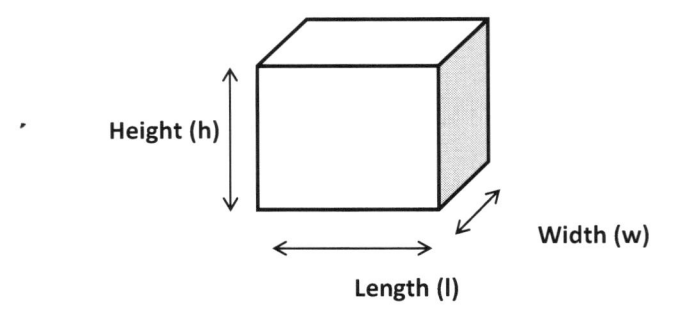

Height (h)

Width (w)

Length (l)

Volume of a cuboid is Height × Length × Width or V = h×l×w (units cubed e.g. cm^3 or m^3, etc)

Example: Find the volume of a box whose dimensions are shown below:

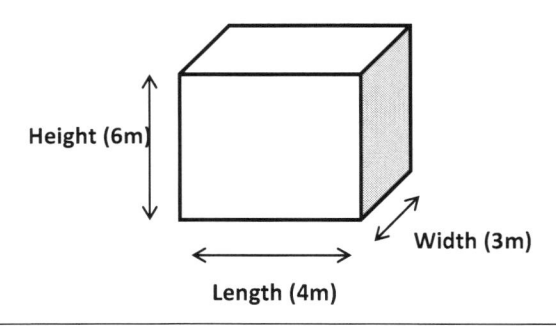

Height (6m)

Width (3m)

Length (4m)

Method: Volume of a cuboid or a box as shown above = Height × Length × Width

So the volume of this box is 6×4×3 = $72m^3$

(Note: volume is measured in cubic units)

Practice questions 9

(1) Find the perimeter of the rectangle whose length is 12cm and width is 9cm as shown below.

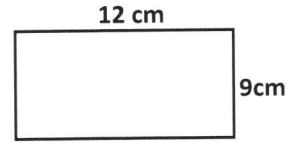

12 cm

9cm

(2) Find the area of the rectangle above

(3) Mary decides to have a flower bed in the middle of her rectangular garden shown below. The flower bed has an area of $12m^2$. The rest of the garden is lawn. What is the area of the lawn?

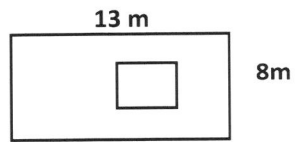

13 m

8m

(4) What is the perimeter (circumference of the circle) whose radius is 7 cm.
(The formula for working out the circumference is C = 2πr) Take the value of π to be 3.142

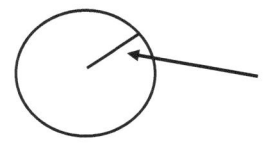

Radius (7cm)

(5) What is the area of the circle above? (The formula for the area of a circle is $πr^2$)

(6) Find the area of the triangle shown below:

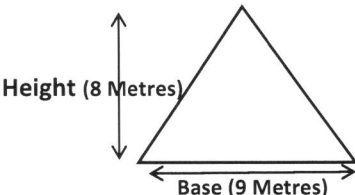

Height (8 Metres)

Base (9 Metres)

(7) John wants to use his box for some books that he has. He works out that on average he can put 16 books in a space of $0.25m^2$. How many books can he put in the box shown below:

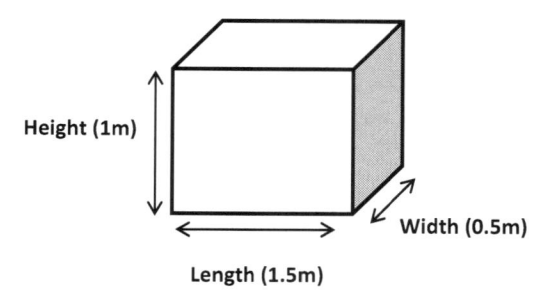

Height (1m)

Width (0.5m)

Length (1.5m)

Answers to Practice Questions 9

(1) 42 cm
(2) 108 cm^2
(3) 92 m^2
(4) 43.99 cm
(5) 153.96 cm^2
(6) 36 m^2
(7) 48 books

2D and 3D shapes and Nets

Common 2 –D Shapes (some of which we saw earlier.)

Square

Rectangle

Triangle

Parallelogram

Rhombus

Trapezium

Pentagon

Hexagon

Octagon

Kite

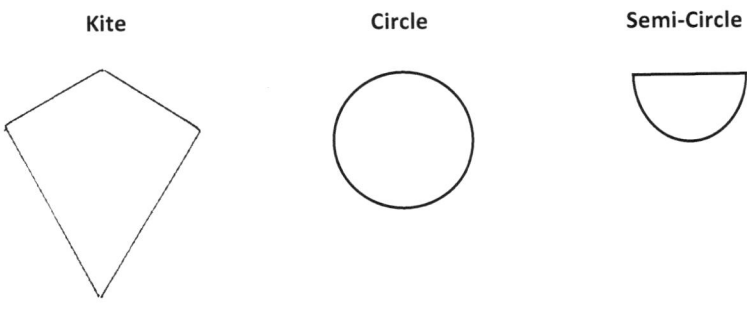

Circle

Semi-Circle

Common 3-D Shapes

Cube

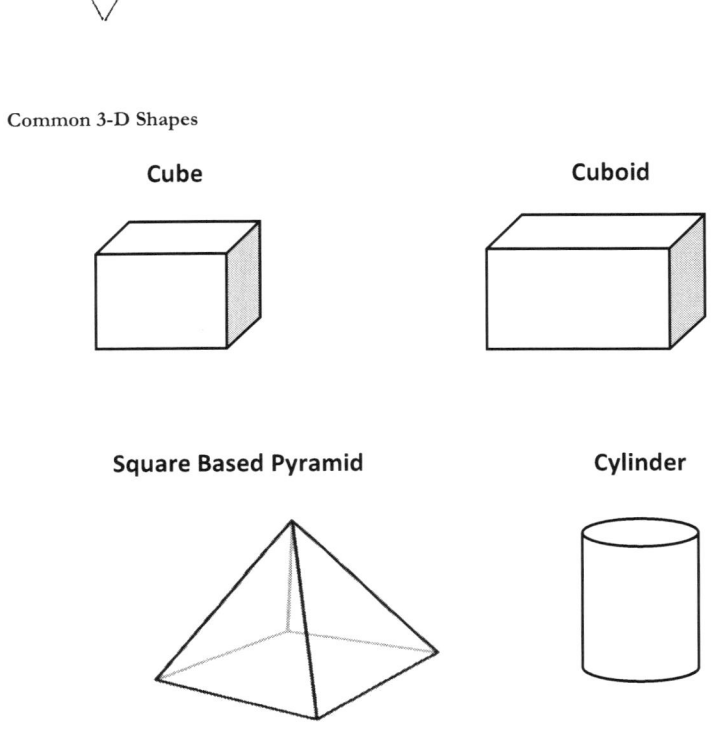

Cuboid

Square Based Pyramid

Cylinder

Nets and their corresponding 3D shapes

When a **3D shape is opened out flat** you should be able to **fold it up again to get the same shape**.

Net of a cube

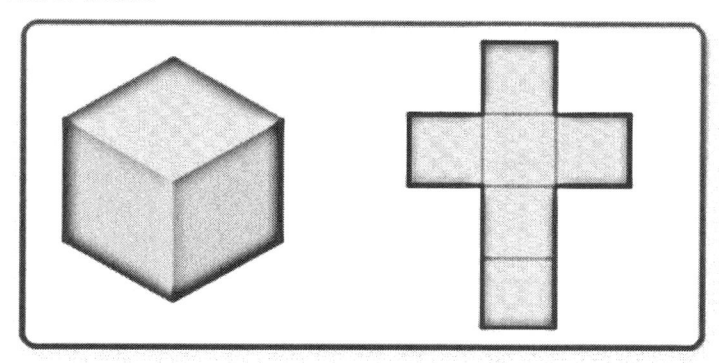

You will find that in a square shaped cube where all the sides are equal the net looks like it is shown on the right. If you fold this up you will get the solid cube again.

Some more examples are shown below:

Net of a cuboid

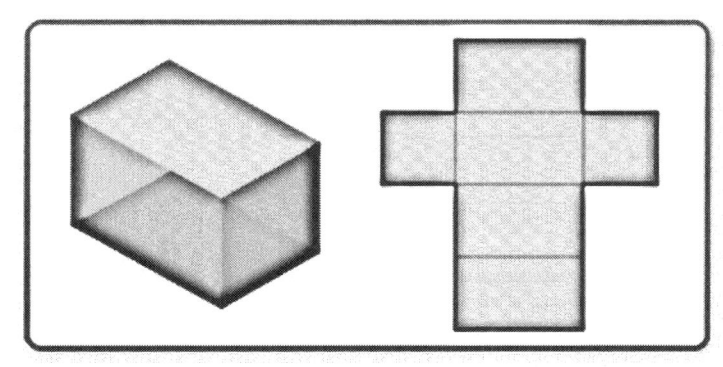

Net of a square-based pyramid

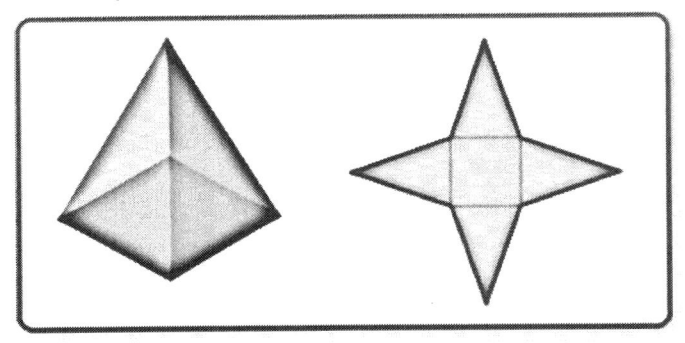

Lines of symmetry.

Some shapes have one line of symmetry others have more and some have none! If you can reflect (or flip) a half of a figure over a given line and that part of the figure remains the same, then the figure has a line symmetry. The line that you reflect over gives you a mirror image of the other half.

As we will see some shapes have more than one line of symmetry and some shapes have no line of symmetry

Examples of shapes with one line of symmetry. (The dotted line is called the line of symmetry.

Example 1:

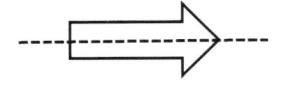

Example 2:

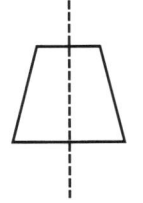

Example 3:

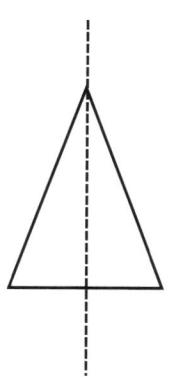

Shapes with two lines of symmetry

Example 1:

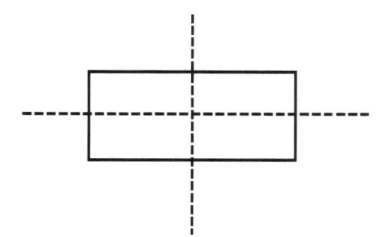

Example 2:

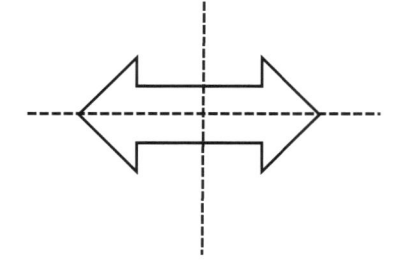

Example of a shape with no lines of symmetry.

You cannot draw a line which will reflect one half of the shape

Plans:

A plan is simply a drawing of what your room, garden or a building will look like.

A simple plan of a garden with a lawn and section for flowers is shown below.

Example 1:

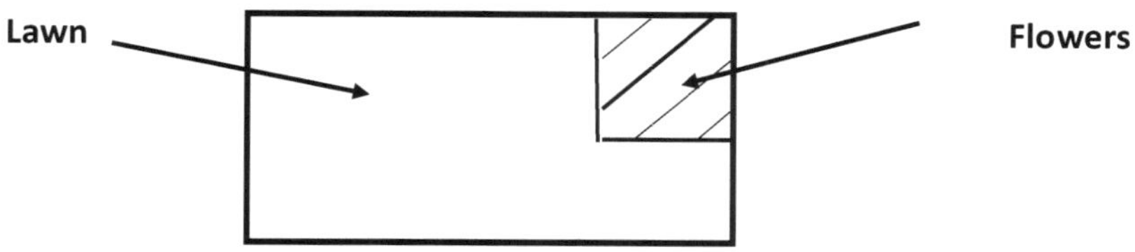

Lawn
Flowers

Example 2:

A 2D plan for example of someone's floor.

REF
KITCHEN
12'-10" x 10'-1"
BATH
LIN
BEDROOM
9'-11" x 11'-4"
REAR ENTRY
MICRO
DW
CL
HALL
CL
CL
LIVING ROOM
15'-9" x 17'-3"
BEDROOM #1
13'-6" x 13'-10"
UP
CL
ENTRY

Example 3:

Jeremy decides a plan for his garden. He wants a small area for a circular pond an area for some flowers and a lawn. The plan is shown below. The area of the pond is $7.1m^2$. The area allocated for the flowers is $10m^2$. What is the area of the lawn?

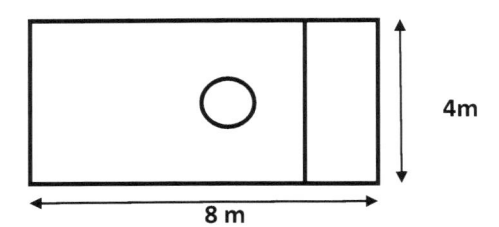

Method: Area of garden = length × breadth = 8×4 = $32m^2$

Since area allocated for flowers is $10m^2$ and the area of the pond is $7.1m^2$

This means the area of the lawn is 32 − 10 − 7.1 = $14.9m^2$

Tessellations

A tessellation is a pattern made by fitting together shapes that leave no gaps.

Example 1: Tessellation with same sized triangles.

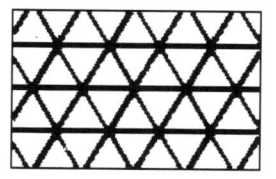

Example 2: Tessellation with same sized squares

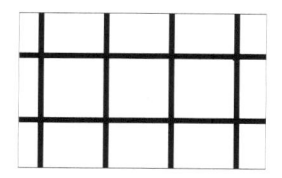

Example 3: Tessellation with a regular polygon

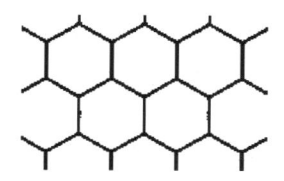

A six sided regular polygon is called a hexagon

Example 4: Tessellation with triangles and squares (Notice they join together in such a way that there are no gaps between the shapes.

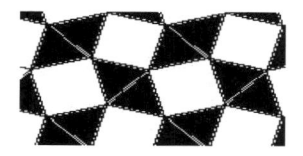

Other Tessellations

You can even have **curved shapes** that join together so long as there are no gaps between the joining shapes.

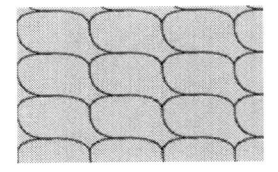

Practice questions 10

(1) What type of a 3D shape can I make from the net below?

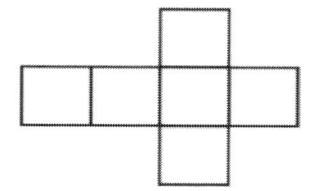

(2) Draw a line of symmetry for the 2D shape below:

(3) Jonathan has a plan for his garden as shown below. The area occupied by the flowers is $4m^2$, the area of the pond is $3.14m^2$, the area of the patio is $20m^2$. Given that the length of the garden is 9m and the width is 4m what is the area occupied by the lawn?

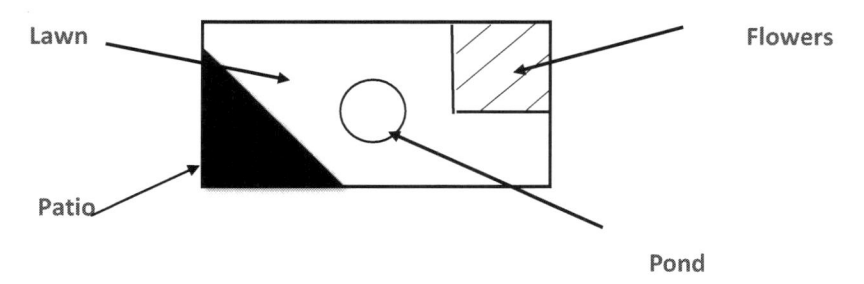

Lawn

Flowers

Patio

Pond

(4) What is meant by a tessellation?

Answers to Practice Questions 10

(1) A cube such as the one shown:

(Note all sides are of equal length)

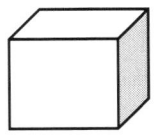

(2)

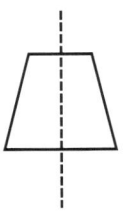

(3) 8.86 m^2

(4) A tessellation is a pattern made by fitting together shapes that leave no gaps

Chapter 10   More Data Interpretation

Mean, Median and Mode

These are simply different types of 'averages'.

Mean: The sum of the numbers in a data set divided by the number of values in the Set

Median: The middle number of a data set when listed in order

Mode: The most frequently occurring number or numbers in a data set

Range: This is the difference between the highest and the smallest numbers in a data set

Example 1:
Find the mean value of the following data set:
2, 7, 1, 1, 7, 8, 9

Method: Find the sum first
2 + 7 + 1 + 1 + 7 + 8 + 9 = 35
Now divide this total by 7, since this is the total number of numbers
So, 35/7 = 5
Hence, the mean value of this data set is 5

Example 2:
Find the median of 3, 7, 1, 8, and 6

Method: First re-order from smallest to biggest, re-writing the numbers we have: 1, 3, 6, 7, 8
Clearly the middle number is 6.
Hence, the median is 6

Example 3:
Find the median of 3, 6, 7, 1, 8 and 5

Method
First re-arrange to get 1, 3, 5, 6, 7, 8
Notice, in this case the middle number is between 5 & 6
So the median is (5 + 6)/2 = 5.5

Example 4:
Find the Range of the data set 3, 5, 7, 1, 8, and 11

Method: Find the difference between the biggest and smallest numbers
So the Range = 11 – 1 = 10

Example 5:
Find the Mode of the following numbers:
1, 4, 4, 4, 7, 8, 9, 9, 11, 12

Method: Find the most frequently occurring number. The most frequently occurring number is 4.
Hence the Mode is 4

Example 6:
Seniors in a care home are encouraged to walk around a bit more every day. They are each given a pedometer which records the number of steps taken. The number of steps achieved by a group of 7 seniors is recorded as shown below:
5000, 1500, 2000, 1400, 1400, 3000 and 4100 steps

   (1) What is the median number of steps in this group?
       To find the median re-arrange the numbers from lowest to highest and find the middle number:
       Re-arranging we have: 1400, 1400, 1500, 2000, 3000, 4100, 5000
       The middle number is 2000 as shown. So the median is 2000 steps.
   (2) What is the range in steps?
       The range is simply the difference between the **highest** and **lowest** steps.
       So the range is 5000 – 1400 = 3600 steps
   (3) What is the mode?
       The mode is the most frequently occurring number which gives us 1400 steps.

Pie Charts

When data is represented in a circle this is called a pie chart. Basically you need to remember that a full circle or 360 degrees represents all the data (or 100% of the data). Half a circle or 180 degrees represents half the data (or 50% of the data), and similarly 25% of the data is represented by 90 degrees or a quarter of a circle. Essentially, each sector or slice of the pie chart shows the proportion of the total data in that category.

Example 1:

The pie chart below shows the percentage of applicants who got different grades in a psychometric aptitude test when applying for a job in a particular company. The requirement to be short listed for a second interview was to pass with high marks. If 140 applicants took this test how many of them were short listed?

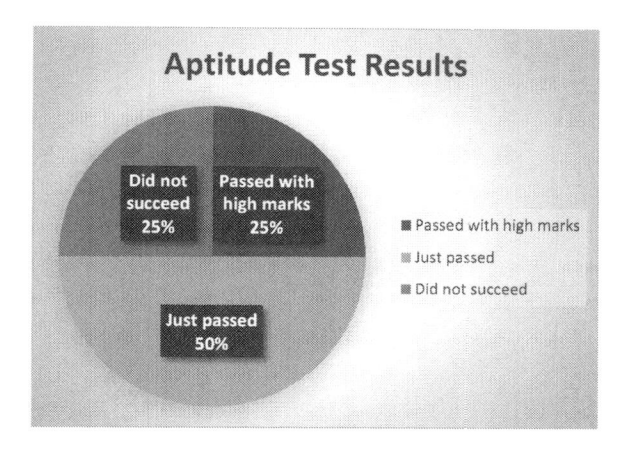

Aptitude Test Results

- Passed with high marks: 25%
- Just passed: 50%
- Did not succeed: 25%

Method: As illustrated the results in this aptitude test for this particular company show that 25% got the required 'high marks' to be short listed for a second interview. Since a quarter of a circle corresponds to 25%. This means a quarter of the 140 applicants attained this which corresponds to 35 people.

Example 2:

The destination of 120 pupils who leave year 11 in School B in 2012 is represented in the pie chart below. The numbers outside the sectors represent the number of pupils

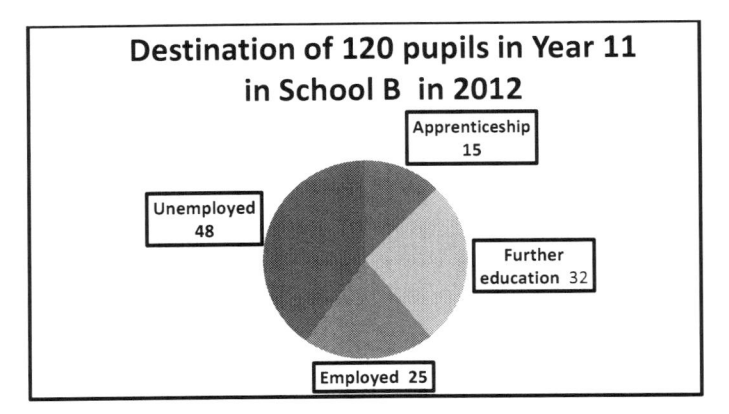

Destination of 120 pupils in Year 11 in School B in 2012

- Apprenticeship 15
- Unemployed 48
- Further education 32
- Employed 25

(1) What is the percentage of pupils who are unemployed?

Method: The number of pupils out of 120 that are unemployed is 48. So the percentage of pupils who are unemployed is $\frac{48}{120}$ × 100 =40%

(**Reminder**: Put 48 in a calculator, divide by 120 and then multiply the answer by 100)

(2) What fraction of pupils go on to Further Education?

Method: The fraction of pupils that go on to further education is $\frac{32}{120} = \frac{16}{60} = \frac{8}{30} = \frac{4}{15}$ the fraction representing this in its simplest form is $\frac{4}{15}$ (Keep dividing the top and bottom of $\frac{32}{120}$ by 2)

(3) What percentage of pupils is either employed or in apprenticeships? Give your answer to one decimal place?

Method: Total number of pupils who are either in employment or apprenticeships = 25+15 =40, hence the percentage is $\frac{40}{120}$ × 100 = 33.3%

(**Reminder:** Put 40 in a calculator, divide by 120 and then multiply the answer by 100)

Bar charts

Bar charts can be represented in columns or as horizontal bars. They can be either simple bar charts that show frequencies associated with data values or they can be multiple bar charts to allow for comparisons between data sets as shown below. The examples below illustrate some of the ways bar charts can be used to represent data.

Example 1: In a cosmetics shop the number of items that were sold for four top brands over a one month period were recorded as shown in the bar chart below.

(1) Which brand had the highest sales? **You can see from the column bar chart below that Brand D had the highest sales as 40 items of this brand were sold during one month, which is higher than any other brand**

(2) What was the proportion of sales for Brand D compared to the total? Give your answer as a fraction in its lowest terms. **The number of Brand A items sold were 20, Brand B were 35 and Brand C were 25 and as we saw earlier 40 items of Brand D were sold. This means the total number of cosmetic items sold during this one month period = 120. Since 40 items belonged to Brand D, compared to the total this is** $\frac{40}{120}$ **which simplifies to** $\frac{1}{3}$

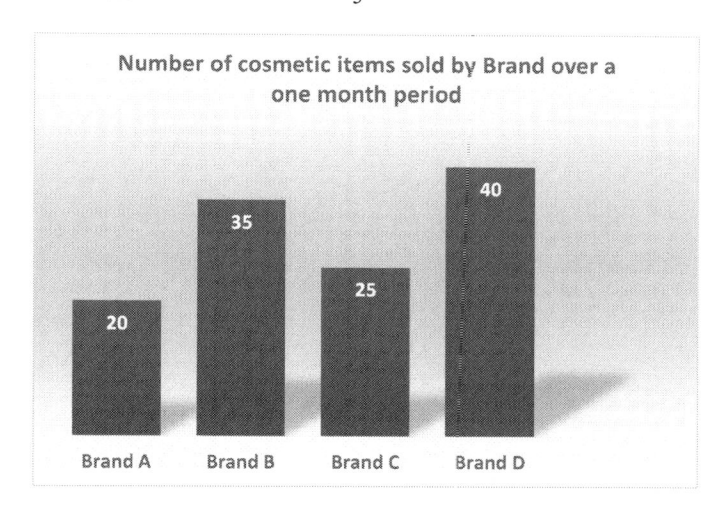

Number of cosmetic items sold by Brand over a one month period

Brand A	Brand B	Brand C	Brand D
20	35	25	40

Example 2:

The bar chart below shows the amount of time in hours John, Bob and Bill spend surfing the web at weekends. What is the mean time per boy that is spent surfing the web at the weekend?

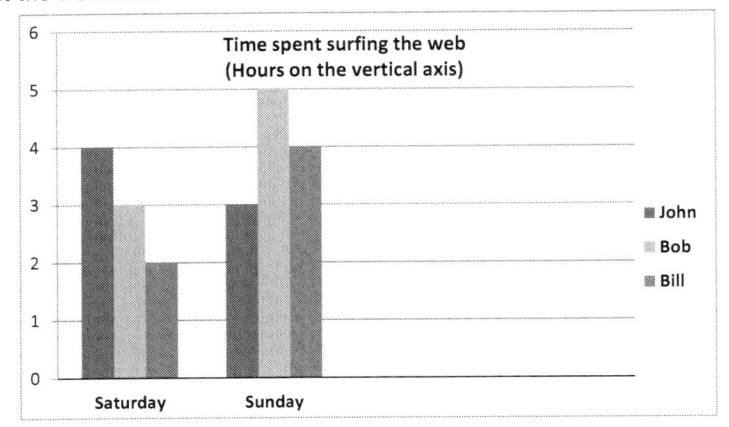

Time spent surfing the web
(Hours on the vertical axis)

Method: John spends 4 hours on Saturday and 3 hours on a Sunday: a total of 7 hours

Bob spends a total of 3 hours on Saturday and 5 hours on Sunday: a total of 8 hours

Similarly, Bill spends a total of 2 + 4 = 6 hours on a weekend

Total time spent surfing between the 3 boys on a week end is 7+ 8 + 6 =21hours

Hence the mean time spent per boy is 21 ÷ 3 =7 hours

Example 3: Horizontal Bar Chart

In an on-line company the percentage of employees in 4 key departments is shown below. (1) If there are 560 employees altogether, how many are in the On-line marketing department? (2) How many more employees are there in On-line marketing compared to Customer Service department?

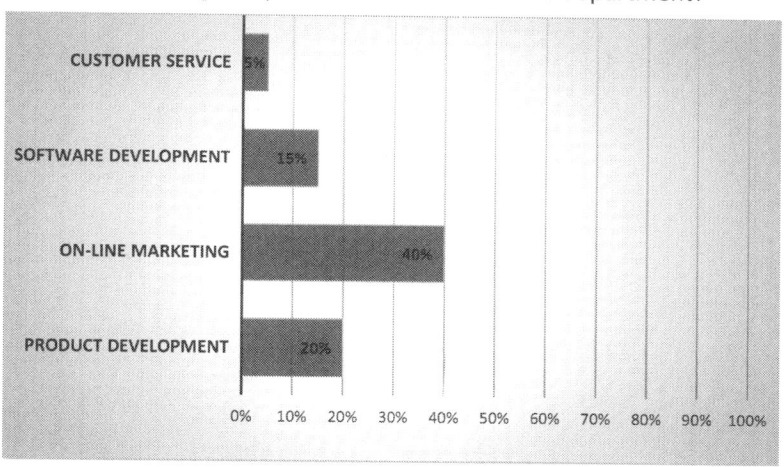

Percentage of employees in different departments in an online company

Method:

(1) From the bar chart you can see that 40% of the employees are in on-line marketing. Since there are 560 employees altogether, this means 40% of 560 = 224 employees

(40% of 560 = $\frac{40}{100}$×560 = 224)

(2) 5% of employees are in customer service. Since there are 560 employees altogether, 10% = 56 and 5% =28 employees. We know from the previous question that there are 224 employees in On-line marketing. So 224 – 28 = 196 employees. This means there are 196 more employees in the On-line marketing department compared to the Customers Service Department.

Example 4:

This **composite bar chart** below shows the percentage of pupils in a particular school who take and do not take additional lessons in music and maths respectively. What percentage of pupils take extra music lessons?

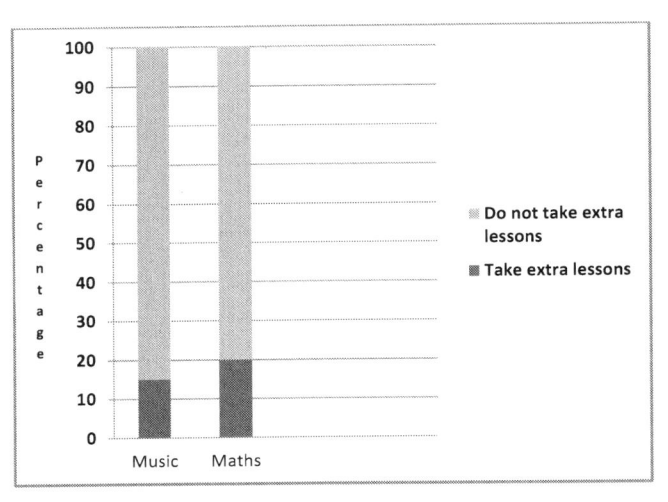

Method: The percentage of pupils who take extra maths lessons is 20%.

Summary of composite bar charts:

Although a composite bar chart consists of single bars, these bars are split into two or more sections. These sections show the frequencies (number of occurrences) of the appropriate categories. Frequency in the above example were the percentages of pupils and the categories in this case were those that took and do not take music and maths lessons respectively.

Line graph

A line graph is a way to represent two sets of related data. **It is sometimes used to show trends**

Example 1: The data below shows the percentage of candidates who had 'A' level mathematics who were short-listed for the first interview when applying to a consultancy company. This data is shown in the table below. However, the same data can be shown as a line graph that follows.

Year	2005	2006	2007	2008	2009	2010
% of candidates with 'A' level Maths	26%	35%	45%	37%	48%	32%

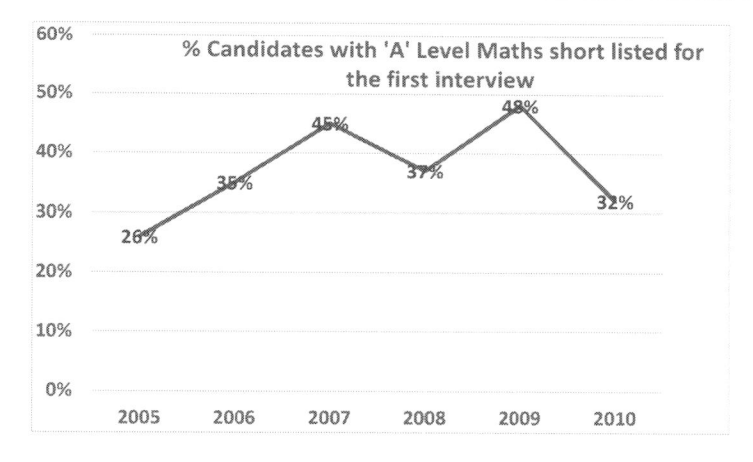

% Candidates with 'A' Level Maths short listed for the first interview

What was the approximate change in percentage points for candidates that were short-listed between 2008 and 2010?

Method: You can see from the table as well as the graph that the success rate actually dropped from (approximately) 37% to (approximately) 32%. That is decreased by 5% points

Example 2:

The sales of Company A and Company B are plotted in a line graph from 2001 to 2006. If 850 employees worked for Company A and 800 employees worked for Company B in 2006. What was the sales per employee for Company B in 2006?

Sales of Company A and Company B from 2001 - 2006 in £ million

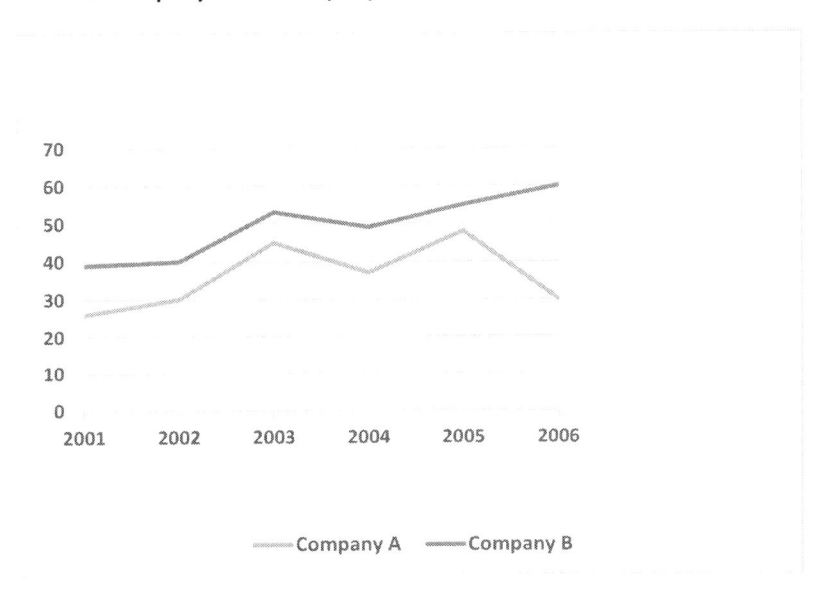

Method: From the line graph it can be seen that in 2006, Company B had a sales of £60 million. If 800 employees worked for this company, then clearly the sales per employee was £60,000,000 ÷ 800 = £75000. Hence, the sales per employee in Company B in 2006 was £75000.

Practice questions 11

(1) What is mean of 3, 7, 8, 2, 9 and 1?
(2) Joseph buys 3 blue candles, 5 green candles, 3 white candles and 7 purple candles. What is the mode?
(3) Find the median of 5, 7, 2, 8, and 4
(4) Find the range of the data set 2, 6, 9, 1, 4, and 15
(5) In the Pie chart below what fraction of students did not succeed?

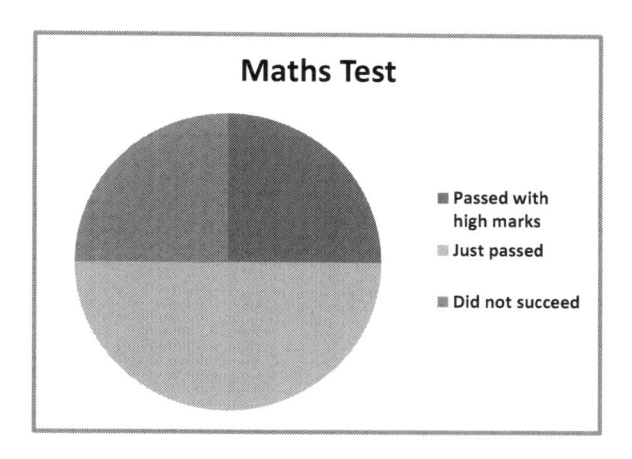

Maths Test
- Passed with high marks
- Just passed
- Did not succeed

(6) The bar chart below shows. How many employees are there in Company D?

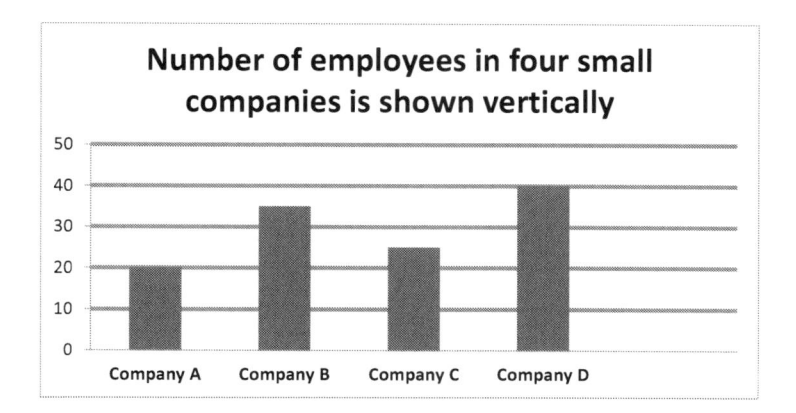

Number of employees in four small companies is shown vertically

- Company A: 20
- Company B: 35
- Company C: 25
- Company D: 40

(7) The bar chart below shows the amount of additional tuition time in hours that Samantha, Joanna and Yasmin spend per week in Maths and English. (a) How many hours of tuition per week does Samantha spend in a week in both subjects? (b) What is mean tuition time spent by the three students in maths?

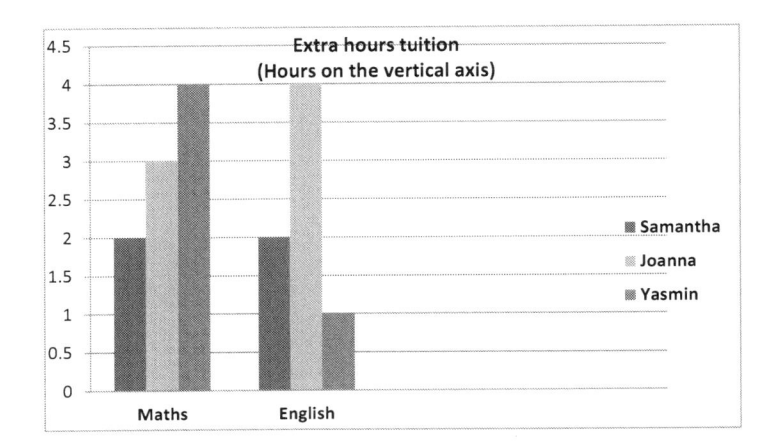

Extra hours tuition
(Hours on the vertical axis)

- Samantha
- Joanna
- Yasmin

Maths: Samantha 2, Joanna 3, Yasmin 4
English: Samantha 2, Joanna 4, Yasmin 1

Answers to Practice Questions 11

(1) 5

(2) 7 purple candles

(3) 5

(4) 14

(5) $\frac{1}{2}$ or 0.5

(6) 40

(7) (a) Samantha spends a total of 4 hours in tuition in a week

(b) 3 hours

Chapter 11   Probability

Probability is the likelihood of an event happening.

Example 1: It is very likely to rain at least one day in April.

Example 2: There is a 50% chance that if a fair coin is thrown it will land heads

Example 3: There is no chance of picking a red ball from a bag containing 6 blue balls

Example 4: It is certain that you will pick up a red ball from a bag containing red balls.

Example 5: I am not completely sure how the economy will perform next year but experts tell me there is a reasonable chance that it will do well. Perhaps 0.6?

Mathematically: **Probability is defined as the likelihood of an event happening. Probability lies between 0 and 1.**

A probability of 0 means that an event will definitely not happen or it is impossible to happen. Likewise a probability of 1 means is certain to happen. Probability is usually expressed as a fraction, a decimal or a percentage.

Consider two simple cases: There are 4 blue balls in a bag. You take out a ball at random. (1) What is the probability that the ball you pick is red? (2) What is the probability that the ball you pick is blue? Although this is a trivial example you can see that in question (1) it is impossible to pick red ball since all the 4 balls in the bag are blue. Hence the probability of picking up a red ball is 0. Similarly, in question (2) the probability that you pick a blue ball is 1. That is you are certain to pick a blue ball, since all the four balls are blue.

Many events of course happen with a probability between 0 & 1. For example a probability of 0.9 would indicate a high chance of an event happening, whereas a probability of 0.1 would imply a low probability of an event happening. The probability of an event happening is defined as:

$$\frac{number\ of\ ways\ in\ which\ the\ event\ can\ happen}{total\ number\ of\ outcomes}$$

Also note that the probability of an event **not happening** is **1 – the probability of an event happening**

Notation used: P(A) means probability of event A happening. Hence probability of event A not happening would be 1 – P(A).

Typical examples:

Example 1:

There are 5 red, 6 green and 7 blue beads in a bag.

(1) You pick a bead at random from the bag. What is the probability of picking a red bead? Answer $P(R) = \frac{5}{18}$ (Reason: there are 18 beads altogether, and 5 of them are red, so the chance or probability of picking a red bead is 5 in 18 or $\frac{5}{18}$). <u>Notice we have used P(R) to mean the probability of picking a red bead.</u>

(2) What is the probability of picking a green or blue bead? Answer $P(G \text{ or } B) = \frac{13}{18}$

Reason: there are 18 beads altogether, and the number of green and blue beads combined total 13. Hence the probability of picking a green or blue bead is 13 in 18 or $\frac{13}{18}$.

(3) What is the probability of not picking a green bead? Answer: P(not G) = 1 – P(G) = 1 - $\frac{6}{18}$ = $\frac{12}{18}$

A simpler way of doing the same problem is to say that since there are 18 beads altogether and 6 of them are green, then this means that 12 are not green, hence the probability of not picking up a green bead is 12 in 18 that is $\frac{12}{18}$. You could of course simplify $\frac{12}{18}$ to $\frac{2}{3}$ (dividing both the top number 12 and bottom number 18, by 6)

Relative Frequencies:

When you do not know the probability exactly you can use an experimental method of relative frequencies to assess an **estimate** of the probability. Let's say you are not sure whether a die is fair or biased. You test it out by throwing it 200 times and get the number 6, fifty times.

Relative frequency of an event is defined as

$$= \frac{Number\ of\ times\ the\ event\ happened}{Total\ number\ of\ trials}$$

In this example, using the formula above the relative frequency $= \frac{50}{200} = \frac{5}{20} = \frac{1}{4} =$ 0.25, it seems that the die is biased as we would expect the number 6 to occur roughly $\frac{1}{6} \times 200$ times which is around 33 times! (Just for interest to be absolutely sure that the result wasn't just a coincidence we may need to repeat the experiment again with a bigger sample and also do something called 'significance' testing to make sure that the die is really biased)

Expected Number

Example 1: If a fair die is thrown 660 times approximately how many threes are we likely to get?

The probability of getting any number when a fair die is thrown is $\frac{1}{6}$. The expected number of threes = P(3) × 660 = $\frac{1}{6}$ × 660 = 110. We would therefore expect the number three to occur 110 times. (Here, P(3) means the probability of getting the number 3)

Example 2: In a certain company, the probability of passing a numerical skills test by any individual is 0.45. If 100 candidates take this test over a year, how many individuals do you expect to pass this test?

Method: 100 × 0.45 = 45. So we would expect 45 candidates to pass this test.

Multiplication law in probability

When you have independent events (that is the outcome of one is not affected by the outcome of the other) then to find the probability of say event A and event B happening we simply multiply the probabilities of A and B together.

Example 1: What is the probability that we will get two sixes when a die is rolled two times?

Method: Probability that we get '6' followed by '6' $= \frac{1}{6} \times \frac{1}{6} = \frac{1}{36}$

Addition law in probability: When two or more events are mutually exclusive (i.e. they cannot occur together), then the probability of A **or** B **or** C happening is simply found by adding the respective probabilities. That is p(A) + p(B) + p(C).

Example 1: there are 6 blue beads, 8 green beads and 15 black beads in a bag. What is the probability of picking either a green or a black bead?

Method: Altogether there are 29 beads. Probability of picking a green bead = $\frac{8}{29}$, similarly the probability of picking a black bead = $\frac{15}{29}$. Hence the probability of picking either a green or black bead is found by adding the two probabilities together. P(Green or Black) = $\frac{8}{29} + \frac{15}{29} = \frac{23}{29}$

Summary:

(1) The probability of an event happening lies between 0 and 1.
(2) Probabilities can be expressed as a fraction, decimal or a percentage
(3) The probability of an event can never exceed 1.
(4) A probability of 0.1 or $\frac{1}{10}$ or 10% means it is not so likely to happen
(5) A probability of 0.9 or $\frac{9}{10}$ or 90% means it is likely to happen
(6) A probability of 0.5 or $\frac{5}{10}$ or 50% means there is an even chance of the event happening
(7) probability of an event not happening is 1 – the probability of an event happening

Practice questions 12

(1) Bob consults a mortgage broker who estimates his chances of getting a Mortgage is 85%. What is the chance that he does not get a mortgage?

(2) Chloe throws a fair coin 550 times. How many times is Chloe expected to get 'Heads'?

(3) The weather forecast says that the chance it will rain some day in April is 0.96. What is the chance that it does not rain in April?

(4) What is the probability of picking a spade from a pack of shuffled cards? Assume there are 52 cards with no jokers.

(5) John throws a fair coin twice. What is the probability that he gets two 'Tails' consecutively?

(6) 32000 teachers take the required Numeracy test in a particular year. The probability of succeeding the first time they take it is 0.88. How many teachers in this particular year pass the test the first time they take it?

(7) There are 6 green marbles and 9 blue marbles in a bag? A marble is picked at random. What is the probability of picking a green marble?

(8) There are 20 beads in a bag. 7 are blue, 9 are red and 4 are purple. A bead is picked at random what is the chance that the bead that is picked is either red or purple?

Answers to Practice Questions 12

(1) 15%

(2) 275

(3) 0.04

(4) $\frac{13}{52}$ or $\frac{1}{4}$ or 0.25 or 25%

(5) $\frac{1}{4}$ or 0.25 or 25%

(6) 28,160

(7) $\frac{6}{15}$ or $\frac{2}{5}$ or 0.4 or 40%

(8) $\frac{13}{20}$ or 0.65 or 65%

Exam Type Questions

(1) Chris decides to print and distribute leaflets for his decorating business. He typically gets 1 customer for every thousand leaflets and each customer gives him an average of 3 days' work. He charges £120 per day. He decides to print and distribute 3000 leaflets. (a) How many days of work can he expect to get?   1 mark

(b) How much in total will he earn after his printing and distribution costs? 3 marks

The cost of printing and distribution is shown below:

Number of leaflets	1000	2000	3000	5000	10000
Cost of printing	£40	£70	£90	£130	£160
Cost of distribution	£50	£100	£130	£200	£340

(2) As Chris wants to grow his business and has had some success previously, he now decides to print and distribute 10000 leaflets. He also gets 10 days of extra work through recommendations. (a) How many days of work in total does he get from leaflet distribution and recommendations?   2 marks

(b) How much in total will he make both from recommendations and leaflet distribution after his printing and distribution costs this time?   3 marks

(3) Alex is tiling two walls of his bathroom. One of the walls is 2.5m wide and 3.7m long. The other wall is 1.8m wide and 3.7m long. He uses square tiles in which the lengths of the sides are 35cm by 35cm. How many tiles will he need to get the two walls tiled? (Give your answer in whole number of tiles)   3 marks

(4) Martha and her husband decide to go to Amsterdam from Birmingham. They decide to take 20kg of luggage each and also have a bit more leg room on their flight. They book their flights online. The costs are shown below:

Cost of return flight to Amsterdam from Birmingham:	£39.50 per person
Cost of seats with extra leg room:	£9.50 per person
Cost of luggage 15 kg or below	£0
Cost of luggage above 15 kg to 20kg	£11.50 per person
****************Online discount: 20%*****************	

How much do they pay in total for their flight, luggage and extra leg room?   5 marks

(5) George is visiting a friend in Exeter. George lives in Manchester. He looks at the map and measures the possible routes to get to Exeter. His measurements on the map show that his best option comes to 27.5cm on the map. He makes a note of the scale which is 1cm for every 6 miles. He also notes that because of some small roads and possible traffic jams he will average 30 miles per hour.

    (a) How far is Exeter from Manchester in miles in the route that George decides to take?   1 mark

    (b) How long does it take him to get to Exeter?   2 marks

    (c) When does he get to Exeter assuming he leaves Manchester at 1330?   3 marks

(6) The graph below shows the percentage of pupils achieving Grade C in Maths from 2006 to 2011.

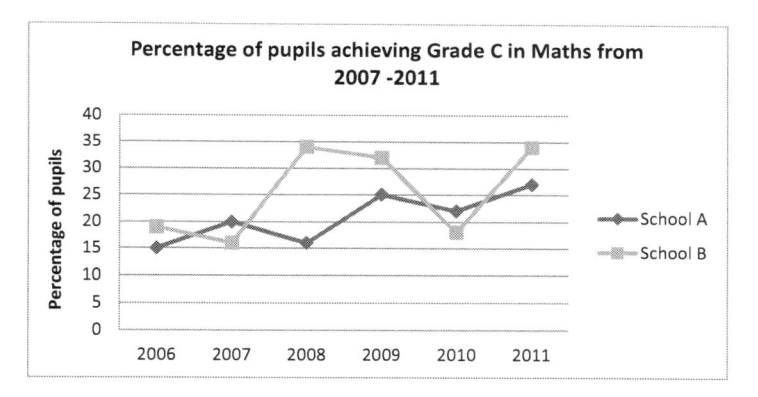

[Graph: Percentage of pupils achieving Grade C in Maths from 2007-2011, showing School A and School B data from 2006 to 2011, y-axis 0-40]

(a) What was the percentage of pupils who achieved a grade C in Maths in 2009 in School A?   2 marks

(b) By how many percentage points approximately did School A outperform School B in 2007   2 marks

(7) The table below shows the total sales by a fashion retailer in London, Paris and New York

Total Sales	2008	2009	2010	2011
London shops (In Millions of £)	10.5	9.8	9.5	10.1
Paris shops (In Millions of Euros)	7.7	7.8	6.9	8.2
New York shops (In Millions of $)	15.1	14.3	14.6	14.9

Assuming that in 2011 on average the exchange rates was £1 = 1.25 Euros and £1 = 1.6 US $. What were the total sales in 2011 for all three cities in pounds sterling? Give your answer in pounds million to two decimal places.

5 marks

(8) A company calculated that it had given bonuses to its junior and senior staff in the ratio of 1:3. There was a total of £68000 bonus given. Assuming there were 20 senior staff, how much did each member of the senior staff get? 3 marks

(9) There are 24 employees in a small company. Three of them go on a special training course. What is the fraction of employees that do not go on this training course? Give your answer as a fraction in its simplest form. 2 marks

(10) Peter belongs to a walking group that walks 24 Km every week. If 8km is approximately equal to 5 miles, estimate how many miles the weekly walk consists of? 2 marks

(11) Joanna is raising funds for a Cancer Charity. She manages to persuade 18 people to give £3.50 each for this charity. What is the total amount she collects?
2 marks

(12) A group activity consists of 16 tasks. Each task lasts 15 minutes. How many hours will this group activity last? 2 marks

(13) Jane has organized a meeting that begins at 10:50. She has planned to start with a general introduction for 6 minutes, a power-point presentation for 18 minutes and finally a question and answer session for 26 minutes. When does the meeting end? Give your answer using the 24-hour clock.

3 marks

(14) In a certain apprenticeship scheme, the probability of an individual being accepted is 0.26. If 50 candidates apply for this apprenticeship, how many individuals are likely to be accepted?
2 marks

(15) Jamie and Susan decide a plan for their garden. They want a small circular pond and an area for some flowers and a lawn. The plan is shown below. The area of the pond is $12.6m^2$. The area allocated for the flowers is $15m^2$. What is the area of the lawn?
4 marks

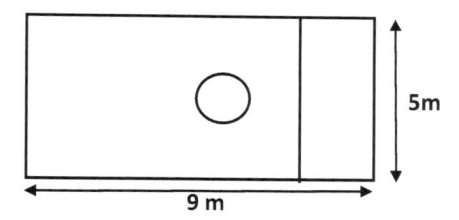

(16) Robin is deciding to tile part of his bathroom. He has a budget of £175. He will tile the part required by himself. He decides to go for white tiles. The area he wishes to tile is $6.25m^2$. He can cut the tiles as needed. The table below shows the prices of small and medium size tiles. (a) Fill in the last column of how many boxes he needs for small and medium sized tiles. (You will need to estimate whole number of tiles and hence the number of boxes required.)
5 marks

(b) Which size will meet his budget?
2 marks

White Tiles	Dimension	Number of tiles per box	Price per box	No of boxes needed
Small Tiles	30cm×40cm	12	£42	
Medium Tiles	40cm×50cm	8	£43	

(17) Susan earns £28,700 as a town planner for a local authority. The new tax legislation allows her a personal allowance of £10,600 before tax is deducted. On this salary her taxable income is 20%. What is her income after the tax she pays? 3 marks

(18) The bar chart below shows the percentage of employees who earn £30,000 per year in Company A. What was the mean percentage of employees who earned £30,000 from 2008 to 2011?

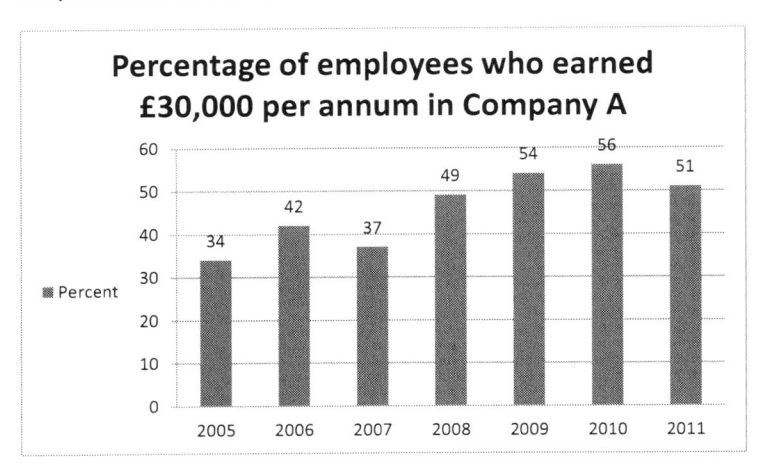

Percentage of employees who earned £30,000 per annum in Company A

Year	2005	2006	2007	2008	2009	2010	2011
Percent	34	42	37	49	54	56	51

(19) An assistant meteorologist wants to convert a temperature of Celsius into Fahrenheit. The formula for converting the temperature from Celsius to Fahrenheit is given by: $F = \frac{9}{5}C + 32$ (where C is the temperature in degrees Celsius). If the temperature is 25 degrees Celsius what is the equivalent temperature in Fahrenheit? 3 marks

(20) Fatima produces hand knitted baby jumpers at home which she then sells on ebay. The costs and selling price of the three types of baby jumpers she makes are given below. What is her percentage profit from the blue jumper she makes? 3 marks

	Cost of Making	Selling Price on ebay
Red Jumper	£2.60	£6.00
Blue Jumper	£3.80	£6.27
Green Jumper	£2.90	£5.50

Answers to Exam Type Questions

(1) (a) Answer: 9 days of work

Method: 3 days of work per 1000 so 9 days for 3000

(b) Answer: £860

Method: Subtract costs from earnings: Total Earnings = 9×120 = £1080

Costs for printing & distributing 3000 leaflets = 90 + 130 = £220

Total earnings after costs = 1080 − 220 = £860

(2) (a) Answer: 40 days of work

Method: Total days = days from recommendations plus 10,000 leaflet distribution. This means: Total days = 10 + 30 = 40 days

(b) Answer: £4300

Method: Total earnings expected from 40 days of work = 40×120 =£4800. Since costs of printing & distributing 10,000 leaflets = £500. This means the total earnings after costs = 4800 − 500 = £4300

(3) Answer: 130 tiles

Method: Area of two walls to be covered are 2.5×3.7 + 1.8×3.7 = 15.91m^2. Area of each tile = $\frac{35}{100} \times \frac{35}{100}$ = 0.1225 m^2 so the number of tiles needed = area to be covered ÷ area of each tile = $\frac{15.91}{0.1225}$ = 130

(4) Answer: £96.80

Method: Total cost of flights =£39.50×2 = £79. Total cost of seats with extra leg room = £9.50×2 = £19. Total cost of luggage = £11.50×2 = £23. Hence total cost of journey before discount = £79 + £19 + £23 = £121

Since the discount is 20% if booked online. We find that 20% of 121 = £24.20. We now subtract £24.20 from £121 to give us £96.80 as the final cost.

(5) (a) Answer: 165 miles

Method: Map distance is 27.5 cm. Since each cm = 6 miles the actual distance is 27.5×6 = 165 miles

(b) Answer: 5.5 hours or 5 hours and 30 minutes

Method: Time taken is distance ÷ speed = 165÷30 = 5.5 hours or 5hrs 30mins

(c) Answer: 19:00

Method: He arrives Exeter 5hrs 30 min after leaving at 1330. This means he arrives at 19:00

(6) (a) Answer: 25%

Method: From the line graph the percentage of pupils who achieved grade C in Maths in 2009 in School A was 25%

(b) Answer: 5%

Method: From the line graph the percentage of pupils achieving the GCSE grade C in 2007 in school B was 15% and in school A it was 20%. So school A outperformed school B by 5% in 2007

(7) Answer: £25.97M

Method: Sales in £ sterling in 2011 were as follows: London: £10.1M

Paris = 8.2 Million Euros, convert to sterling, 8.2÷1.25 = £6.56M

New York = 14.9 Million dollars, convert to sterling 14.9÷1.6 =£9.3125M. Adding up London, Paris & New York we get: 10.1 + 6.56 + 9.3125 = £25.97M (to two decimal places)

(8) Answer: £2550

Method: Total parts is 4. Each part is worth £68000÷4 = £17000. Senior staff get 3 parts =3×17000 = £51000. Since there are 20 members in the senior staff this means each member of the senior staff gets 51000÷20 = £2550

(9) Answer: $\frac{7}{8}$

Method: 21 out of 24 do not go on a training course. This as a fraction is $\frac{21}{24}$. If you divide top and bottom by 6 this simplifies to $\frac{7}{8}$

(10) Answer: 15 miles

Method: since 8km = 5 miles, then 24km = $24 \times \frac{5}{8}$ = 15miles

(11) Answer: £63

Method: Total collected is 18×3.50 = £63

(12) Answer: 4 hours
Method: Each task lasts 15 minutes or 0.25 hours so 16 tasks last 0.25×16 = 4 hours

(13) Answer: 11: 40
Method: Total time = general introduction for 6 minutes, a power-point presentation for 18 minutes and finally a question and answer session for 26 minutes = 6 + 18 +26 = 50minutes. Start time is 10:50 so finish time is 10:50 + 50 = 11: 40

(14) Answer: 13 candidates
Method: Probability of being accepted ×No. of applicants= 0.26×50 =13

(15) Answer: Area of lawn = $17.4m^2$
Method: Total area = 9×5 = $45m^2$. Area allocated for pond = $12.6m^2$ and area allocated for flower bed = $15m^2$. So area of lawn = 45 – 12.6 – 15 = $17.4m^2$

(16) Answer: (a) 5 boxes of small tiles, 4 boxes of medium sized tiles. As shown below

White Tiles	Dimension	Number of tiles per box	Price per box	No of boxes needed
Small Tiles	30cm×40cm	12	£42	5
Medium Tiles	40cm×50cm	8	£43	4

Method: Each small tile is $\frac{30}{100}$ m by $\frac{40}{100}$ m. So the area of each small tile is 0.3×0.4 = $0.12m^2$. To find the number of tiles required divide the area to be covered ($6.25m^2$) by the area of a small tile. That is 6.25÷0.12 = 52.08 tiles. Which means we need to get 53 tiles. Since each box of small tiles contains 12 tiles. We need to get 53÷12 = 4.42 boxes. But since you can't buy 4.42 boxes we need to buy <u>5 boxes</u>. Similarly, for the medium size tiles, the area of each tile is 0.4×0.5 =$0.2m^2$. This means we will need 6.25÷0.2 tiles = 31.25 tiles, or 32 tiles.
So the number of boxes required if we get medium size tiles will be 4.

(b) Answer: 4 boxes of medium sized tiles will be within Robin's budget. Method: 5 boxes of small tiles cost 42×5 = £210 and 4 boxes of medium sized tiles cost 43×4 = £172. So Robin needs to buy 4 boxes of medium sized tiles which is within his budget of £175

(17) Answer: £25,080 after tax
Method: Taxable income is 28,700 – 10,600 = £18,100
Since the tax payable on her income is 20% this amounts to
$\frac{20}{100}$×18,100 = £3620. To find her income after tax we need to find
28,700 – 3620 = £25080

(18) Answer: 52.5%

Method: The mean % of those that earned £30,000 from 2008 to 2011 =

(49 54 + 56 + 51)/4 = 210÷4 =52.5%

(19) Answer: 77°F

Method: The formula given for converting the temperature from Celsius to Fahrenheit is given by: F= $\frac{9}{5}$ C +32. Since C = 25, we substitute this value into the formula to give us F = $\frac{9}{5}$ ×25 +32 = 45 + 32 = 77°F

(20) Answer: 65%

Method: Blue jumper costs £3.80 to make and Fatima sells it for £6.27. Her profit is 6.27 – 3.80 = £2.47. So the percentage profit is $\frac{2.47}{3.80}$×100 = 65%

Some Useful Definitions and Reminders

Natural Numbers: are {1, 2, 3, 4, ...}

Whole Numbers: are {0, 1, 2, 3, ...}

Integers: These are whole numbers that include both positive and negative numbers including 0. So for example ....-5,-4,-3,-2, 0, 1, 2, 3, 4, ... are all integers.

Multiples: These are simply numbers in the multiplication tables.

For example the multiples of 6 are 6, 12, 18, 24, 30, ...

Factors: A factor is a number that divides exactly into another number as for example, the number 2 in the case of even numbers.

3 is a factor of 9, as 3 goes exactly into 9. Other factors of 9 are 1 and 9.

15, has two factors other than 15 and 1. The two factors are 5 and 3, since both these numbers go exactly into 15. **Example:** Find all the factors of 21. The factors are: 1, 3, 7 and 21 (since all these numbers divide exactly into 21)

Prime numbers: A prime number is a natural number that can be divided only by itself and by 1 (without a remainder). For example, 11 can be divided only by 1 and by 11. Prime numbers are whole numbers greater than 1. So for example the first 10 prime numbers are: 2, 3, 5, 7, 11, 13, 17, 19, 23 and 29. **Be careful that an odd number is not necessarily a prime number.** For example **9 is not a prime number** as its factors are 1, 3 and 9 and **prime numbers should have only two factors, 1 and the number itself. Also, note that 2 is a prime number, the only even number that can be divided by 1 and itself!**

Lowest Common Multiple (LCM)
This is essentially the smallest number that will divide exactly by the numbers given. Consider the examples below:

Example 1: Find the LCM of 15 and 45

One method is to find the multiples of both numbers and identify the lowest common multiple as shown below:

Multiples of 15 = 15, 30, **45**, 60, 90, ...

Multiples of 45 = **45**, 90, 135, 180, ...

Clearly **45** (the highlighted number above) is the smallest number that is divisible by 15 and 45.

Example 2: Find the LCM of 10 & 15

First find the multiples of each number:

Multiples of 10 = 10, 20, **30**, 40, 50, 60, 70,......

Multiples of 15 = 15, **30,** 45, 60, ...

You can see that **30** is the lowest common multiple since it is divisible both by 10 & 15.

Highest Common Factor (HCF)

This is the biggest number that will divide exactly into all the numbers given

Example 1: Find the HCF of 15 & 45

Method: Find the factors of each number given and then identify the biggest number that will divide into both these numbers as shown below:

Factors of 15 ={1, 3, 5, **15**}, Factors of 45 ={1, 3, 5, 9, **15**, 45}

You can see that **15** is the **highest common factor** which divides into **both** 15 and 45 exactly.

Example 2: Find the HCF of 8 and 32.

First find the factors of each number given

Factors of 8 ={1, 2, 4, **8**}, Factors of 32 = {1, 2, 4, **8**, 16, 32}

You can see that the number 8 is the highest **common** factor which divides into 8 and 32 exactly.

Square numbers and square roots

Squaring a number is simply multiplying a number by itself.

So 4^2 means 4 × 4 =16, 12^2 means 12 × 12 =144 and so on.

The square root is written like this $\sqrt{}$ and means finding a number which when multiplied by itself gives you the number inside the square root.

Example1: Find $\sqrt{16}$. The answer is clearly 4. Since 4×4 =16

Let us consider some other square roots.

$\sqrt{49} = 7$, $\sqrt{121} = 11$, $\sqrt{100} = 10$, $\sqrt{225} = 15$,

$\sqrt{256} = 16$, $\sqrt{324} = 18$

Cubes

Cubing a number is simply multiplying the number by itself three consecutive times. A cube of a number is written as x^3, where x is the number.

So, for example, 5^3 means 5×5×5 =25 × 5 =125

Similarly, $6^3 = 6 \times 6 \times 6 = 216$, $7^3 = 7 \times 7 \times 7 = 343$, $9^3 = 9 \times 9 \times 9 = 729$,

$10^3 = 10 \times 10 \times 10 = 1000$

Cube Roots

Cube roots are found by finding a number which when cubed gives you the number inside the cube root.

So for example the cube root of 125 is written as $\sqrt[3]{125}$

Also we know that 5X5X5 =125, so that $\sqrt[3]{125} = 5$

Printed in Great Britain
by Amazon